Marie-Hélène LE DU,
Pierre LEGRAND, Serena SIRIGU
and Sylvain RAVY

Preface by Dr Anne Houdusse from the French Academy of Sciences

Introduction to Biological Crystallography

Science Press | edp sciences

Printed in France

EDP Sciences – ISBN(print): 978-2-7598-3501-0 – ISBN(ebook): 978-2-7598-3502-7
DOI: 10.1051/978-2-7598-3501-0

Preamble

The use of three-dimensional structures of biological macromolecules is part of the daily life of many scientists. These structures make it possible to understand how molecules work, to design mutants to study their function or to design drugs to modulate their activity. However, the reliability of a three-dimensional structure is not always the same. Many parameters can be used to assess this reliability. To make the best use of a macromolecular model, it is essential to maintain a critical perspective, which requires knowledge of the strengths and limitations of the method used to construct it.

In order to provide biologists with the necessary basis for this critical view, we developed the MOOC "Voyage au cœur du vivant avec des rayons X : la cristallographie" with the help of many colleagues. Several broadcast sessions were held on the FUN-MOOC platform (https://www.fun-mooc.fr/), as well as some private sessions (SPOC: Small Private Online Course) to accompany training workshops in biological crystallography. A MOOC is designed to be completed entirely online at a given time. This temporal aspect allows a more or less live exchange between students and tutors in a dedicated forum. The disadvantage is that the course content is not accessible outside the MOOC sessions, which can also be limiting.

To complement the MOOC, we decided to write this book, derived from its content, revised and enriched so that it can be used independently. Like the MOOC, this book is an introduction to biological crystallography. It is aimed primarily at biologists, but also at anyone interested in structural biology. Our starting point is the history of this century-old and multidisciplinary method. In addition, we have included 'boxes' within each chapter that cover certain basics or elaborate on certain points in order to satisfy different reading levels. At the end of each chapter, we have grouped the links and QR codes that give access to the MOOC videos corresponding to the chapter in question. Finally, we have grouped references in the 'Further Reading' appendix.

DOI: 10.1051/978-2-7598-3501-0.c901

The diversity of our scientific backgrounds is reflected in this eight-handed manuscript, in which we have tried to maintain a style that is accessible to as many people as possible.

This work would not have been possible without the support of the Atomic Energy and Alternative Energies Commission (CEA), the University of Paris-Saclay, the SOLEIL synchrotron, and the National Centre for Scientific Research (CNRS).

Finally, we would like to express our sincere thanks to Dr Anne Houdusse of the French Academy of Sciences, who gave us the honour of writing the book's preface.

Marie-Hélène LE DU is a CEA research director at the Institute of Integrative Cell Biology (I2BC/CNRS) at the University of Paris-Saclay.

Pierre LEGRAND is in charge of the Proxima-1 beamline at the SOLEIL synchrotron.

Serena SIRIGU is a beamline scientist on the Proxima-2A beamline at the SOLEIL synchrotron.

Sylvain RAVY is a CNRS research director at the Solid State Physics Laboratory (LPS/CNRS) at the University of Paris-Saclay.

Preface

"A great advantage of X-ray analysis as a method of chemical structure analysis is its power to show some totally unexpected and surprising structure with, at the same time, complete certainty."

Dorothy Hodgkin

In our post-genomic era where the acquisition of scientific data continues to accelerate, a major challenge is to identify and validate key interactions underlying particular cellular processes. The interactions between cellular components are at the foundation of life processes and cannot be easily predicted. Functional and structural study of these components is essential to the decryption of such interactions and is critical to guiding the experiments of cell biologists and microbiologists. Thanks to molecular biology and the visualization of these interactions, cell biology experiments can be designed more precisely to study the impact of mutations or interactions on cellular processes of interest. The fundamental mechanisms that govern life can thus be deciphered.

Discovering the processes of life requires diving into the world of macromolecules, the principal components of living organisms. Decisions within a cell are governed by the multiple interactions and chemical mechanisms of which these components are capable. To account for the functions of these macromolecules and the complexes that they can form, it is essential to study their three-dimensional structure, that is to say, the precise positioning of the atoms of these cellular components in space. Indeed, this knowledge provides three-dimensional visualization of interactions between cellular components and is essential to accessing their dynamic properties.

Several three-dimensional structures are often required to interpret the function of a macromolecule. Often, the only way to access mechanistic information is to report on changes in structure at atomic resolution when cellular partners recognize each other or undergo a chemical reaction. In such cases a so-called conformational change often takes place. Understanding the structure and dynamics of macromolecules provides essential clues towards an understanding of cellular processes and

DOI: 10.1051/978-2-7598-3501-0.c902

how they are turned on and off. Moreover, the interaction with small chemical molecules and drug candidates can modulate these cellular processes: the cellular function is either inhibited or enhanced for the benefit of patients. Obtaining a three-dimensional structure with a drug candidate bound directly reveals the molecular interaction and can therefore guide the synthesis of more effective drug candidates. Structural biology is critical to understanding the mechanisms of life at the most detailed and precise level while also directing research into new pharmacological therapies.

With 23 Nobel Prizes awarded in the field, crystallography has provided powerful insights for biologists throughout the 20th century. Many challenges were met with increasing expertise and additional computing power, together with the development of exquisite light sources at synchrotrons. The development of this discipline is an edifying example of cooperation for the benefit of life sciences, in particular through the CCP4 collaborative computing project. This project facilitated the mobilization of both users and developers to work together on computation approaches, which made it possible to solve three-dimensional structures with continuously increasing efficiency. The first chapter of this book relays the history of this discipline. Throughout the chapter, a sense of humility and gratitude emerges in the face of the ingenuity of our pioneers who have enabled methodological and technological leaps forward. In particular, the crystallography community owes much to the scientists who contributed to the development of large instrument beamlines. Today, crystallography is a robust, efficient, and accessible method for visualizing the constituents of life at high resolution.

The following chapters present the different steps and methodologies required to solve a structure by crystallography. They are written in a didactic manner while being complete and pragmatic. The clear presentation of theoretical principles is combined with practical considerations for carrying out experiments at each stage of structure determination. The content is tailored to demystify biological crystallography making it accessible to any scientist. The authors of this book thus provide a valuable document for fledgling researchers in structural biology by allowing them to understand the principles that make crystallographic structure determination possible. The important parameters for evaluating the quality of the structural data available in the Protein Data Bank (PDB) are also defined, notably in chapter 13.

Thorough knowledge of macromolecular atomic structure is essential to understanding the mechanisms of life. The wealth of information in a molecular structure lends clarity to the results obtained by other disciplines of biology. By producing verifiable hypotheses, a structure is a valuable guide for designing functional cell biology or *in vitro* reconstitution experiments aimed at describing the regulation of life processes and their dysfunction in the event of disease or invasion by a microorganism or parasite. This book provides critical information which grants access to the basic principles of structure determination. As such, this book encourages biologists to take into account the wealth of information contained in a three-dimensional structure and will help to bring together biological, molecular, and cellular biology approaches, whose cooperation is essential.

As current technological revolutions allow young researchers to acquire data more and more rapidly, this should inspire them to test increasingly precise mechanistic questions *via in vitro* reconstitution and *in cellulo* assays. Innovative studies questioning how a cell's environment or its interactions with other cells influence internal cellular processes will broadly describe how a cell is sensitive to its surroundings. The challenge however is to decipher the cellular mechanisms that read and integrate this information. By visualizing the molecules of life, we create a gateway rich in information that has been amassed at the PDB. Biologists must work together to exploit this information and formulate hypotheses that will guide cell biology, biochemistry, and structural biology experiments. We will thus be able to fully understand the conformational changes of the macromolecules that underlie the processes of living things, whether physiological or pathological.

"The important thing in science is not so much to obtain new facts as to discover new ways of thinking about them."

William Lawrence Bragg

Anne Houdusse

Anne Houdusse is a CNRS research director at the Cell Biology and Cancer department at the Institut Curie, Paris, France.

Contents

Table of Inserts

Introduction

What are biological macromolecules?

Living tissues are made up of organic molecules, some of which can assemble into molecules of tens of thousands of atoms called macromolecules (see box on **The different types of macromolecules**). Depending on their location and structure, these macromolecules fulfil a variety of essential functions. To understand the function of a macromolecule, it is essential to know its three-dimensional structure, *i.e.* the spatial organisation of its atoms. Such information allows us to understand the mechanisms behind the functions of living organisms. Beyond understanding, it can provide the basis for designing molecules that either block, activate or mimic these functions.

The different types of macromolecules

In biochemistry, there are three main types of macromolecules, which are biopolymers formed by the covalent binding of a few hundred to a few thousand blocks of molecules:

- Polysaccharides, or complex sugars, are made up of monosaccharides. Each monosaccharide contains 5 or 6 carbon atoms, such as glucose (6 carbon atoms), fructose or ribose (5 carbon atoms), naturally folded into a ring. Their purpose is to store energy (starch in cereals) or to structure cells (cellulose in wood).
- Nucleic acids are made up of nucleotides. Each nucleotide consists of a nitrogenous base (purine or pyrimidine), a sugar (ribose or deoxyribose) and a phosphorus group. The nature of the sugar determines the category of the nucleic acid: ribonucleic acid (RNA) if it is a ribose, deoxyribonucleic acid (DNA) if it is a deoxyribose. Their role is to store and transmit genetic information, to regulate gene expression, and sometimes to catalyse it.
- Proteins are made up of amino acids linked in sequence to form a peptide chain (see box, chapter 2). Their functions can range from catalysis to movement, to signalling, and so on.

A key aspect of the functions associated with macromolecules is the number of stable or transient interactions they form with other macromolecules, ligands or organic cofactors, ions. A functional molecular unit can also consist of a stable assembly of macromolecules of different categories: this is the case of the ribosome (proteins + RNA) or the nucleosomes (DNA + proteins). Some macromolecules are themselves made up of a combination of blocks from different categories, such as peptidoglycan (sugars + amino acids) from the wall of certain bacteria. Post-translational modifications correspond to the addition of blocks from another category to a given macromolecule, such as the addition of sugars to proteins – protein glycosylation – or the addition of a methyl group to DNA – DNA methylation.

In the Protein Data Bank (PDB), the international storage database for the structures of biological macromolecules, more than 90% of the deposited structures correspond to single proteins, about 5% to protein/nucleic acid complexes, and 2% to single nucleic acids (https://www.rcsb.org/stats/summary). Because of their great flexibility, polysaccharides are almost absent from the PDB, unless they are in a conformationally stable form such as glycosylation or in complex with a protein. Nucleic acids are also very flexible, although they can adopt stable conformations such as transfer RNA or DNA quadruplexes. Finally, while the majority of proteins adopt a structured fold, many proteins (between 15% and 50%, depending on the organism) or protein fragments, are unstructured – they are intrinsically disordered. Their resulting flexibility is adapted to perform specific biological functions.

DOI: 10.1051/978-2-7598-3501-0.c903

The structure of a macromolecule consists of distinct molecular blocks whose assembly is described on five levels: the sequence (1D), the secondary structure (2D), the spatial arrangement (3D), the molecular assembly (4D), and the dynamics (5D). Thus, for proteins:

- The sequence corresponds to the order of the amino acids that constitute the polypeptide chain with a backbone, a peptide chain, and side chains that characterise each amino acid.
- The secondary structure is the result of interactions within the backbone. It consists, for example, of beta strands or alpha helices.
- The tertiary structure is the result of both backbone and side-chain interactions. It corresponds to the organisation of the sequence and secondary structures in three-dimensional space.
- Molecular assembly is the result of interactions between the polypeptide chain and/or other macromolecules. It involves several polypeptide chains or one or more polypeptide chains and one or more macromolecules of different nature.
- The dynamics are due to the interaction between the protein and the chemical environment. It involves the idea of structural movements over time, changes in the conformation or the shape of a molecule or a molecular assembly (figure 1).

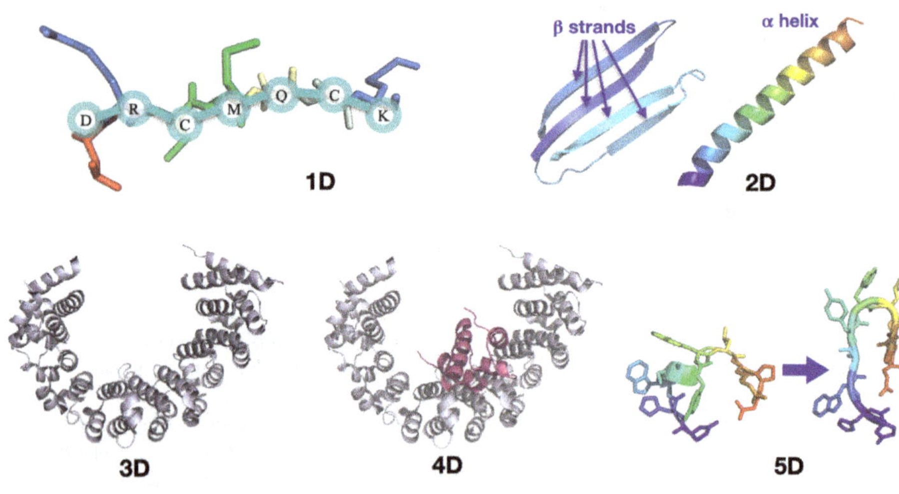

FIG. 1 – Representation of the different levels of structural organisation of a biological macro-molecule.

In structural biology, we are primarily interested in the three-dimensional and quaternary structure of macromolecules, and in their dynamics, as far as this can be investigated. However, not all the macromolecules are necessarily structured (see box on **The different types of macromolecules**).

Let's take the example of a protein. We can visualise it as a ribbon, to highlight the secondary structures. Alternatively, we could choose to visualise only the surface of the molecule to highlight its shape and perhaps the electrostatic potentials generated by the positive and negative charges distributed on the surface. It may also be interesting to represent all the atoms, either as sticks or as spheres whose radius is proportional to the number of electrons in the atom (figure 2). Both representations are based on the same three-dimensional model, but each of them highlights specific properties of the molecule.

Regardless of the representation used to visualise the protein, suitable software reads the coordinates file, which contains the *xyz* position of each atom in the molecule. These files use an international standard, the most widely used being that of the *Protein Data Bank*. These coordinate files are available on two main websites:

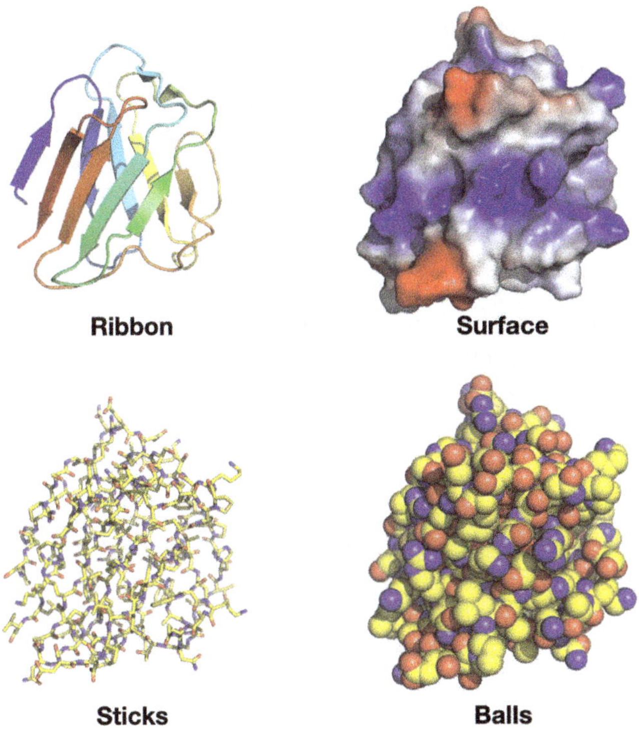

FIG. 2 – Different representations of a biological macromolecule. The arrows on the ribbon representation show the direction of the beta strands. The blue and red colours in the surface representation correspond to positive and negative electrostatic potential, respectively. On the stick or ball representations, the carbon atoms are yellow, the oxygen atoms are red and the nitrogen atoms are blue. Note that hydrogen atoms are not shown as they are generally not visible in crystal structures.

- RCSB (Research Collaboratory for Structural Bioinformatics): http://www.rcsb.org/pdb/home/home.do;
- PDBe (Protein Data Bank in Europe): http://www.ebi.ac.uk/pdbe/.

How to study the structure of a biological macromolecule?

What experimental data and methods allow us to build three-dimensional models to study biological macromolecules?

The scale used to study a macromolecule is the ångstrom (Å), which is 0.1 nm, or 10^{-10} m. Three experimental methods provide three-dimensional information about macromolecules at the atomic scale: cryo-electron microscopy, X-ray crystallography, and nuclear magnetic resonance (NMR) (figure 3). Cryo-electron microscopy requires only a small amount of material to perform the experiments, ranging from a few hundred micrograms to a few milligrams, which is a great advantage when the macromolecule of interest is difficult to produce (see chapter 2). A second advantage is that large molecular assemblies can be addressed since there is no limit to the sample size. On the other hand, electrons damage the sample very quickly, which is why it has to be frozen (figure 3, left). X-ray crystallography offers unprecedented accuracy in determining the position of atoms, and a high degree of robustness and reliability of the result. It has been used to solve the structure of very large macromolecular assemblies, such as whole viruses, the ribosome or the protein synthesis machinery, consisting of more than 100 000 atoms. The sample size limit depends on the conditions under which the diffraction data are recorded (see chapter 6). On the other hand, this method requires a large amount of pure protein (purity >95%), ranging from one milligram to several tens of milligrams. The main bottleneck of this technique is protein crystallisation, which is often the reason why scientists give up crystallography (figure 3, centre). Finally, NMR has the

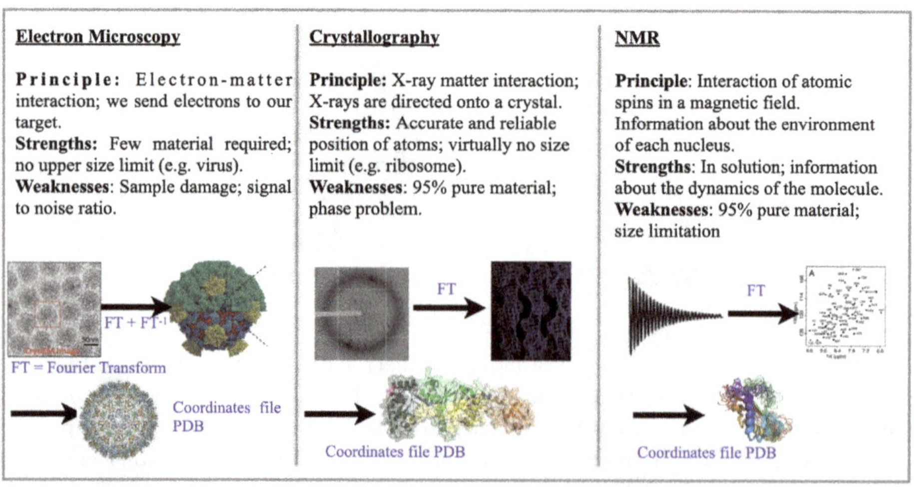

FIG. 3 – General principles, advantages and disadvantages of the three experimental methods for determining the three-dimensional structure of biological macromolecules.

advantage of providing access to the structure of the molecule in solution and can provide information about the dynamics of the macromolecule. However, it is less accurate than crystallography in terms of atomic position, and the size of the molecules that can be targeted rarely exceeds 30 kDa (figure 3, right).

Crystallography remains the most commonly used technique, although cryo-electron microscopy is now being used to study large assemblies. The strength of NMR lies mainly in its ability to provide information about interactions or conformational changes of the molecule in a given cellular environment (figure 4).

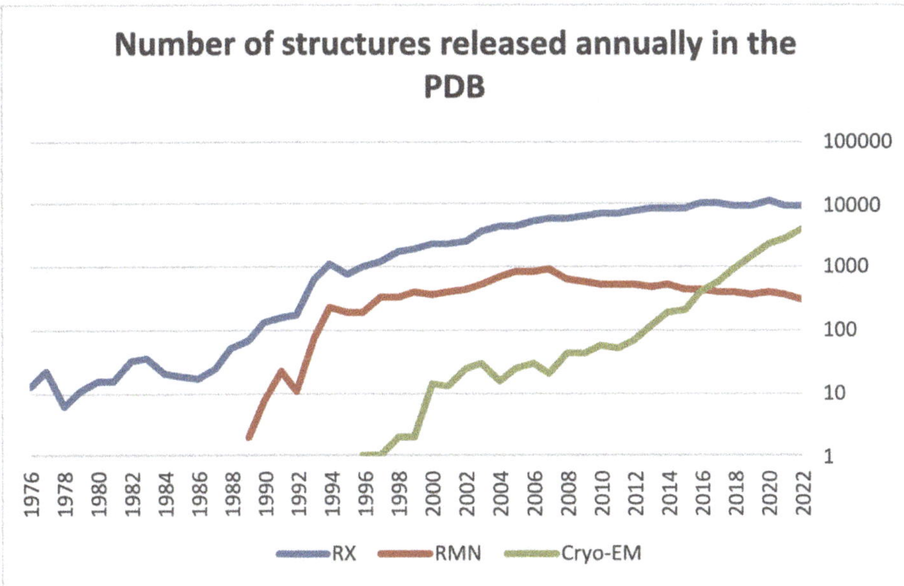

FIG. 4 – Number of structures determined by X-ray crystallography (XRD), nuclear magnetic resonance (NMR), electron microscopy or electron crystallography or electron tomography (EM).

Related videos

What is a three-dimensional structure? (Qu'est-ce qu'une structure tridimensionnelle?)
https://youtu.be/YyqYkiPcyGI

Structural biology approaches (Les méthodes de biologie structurale)
https://youtu.be/BTgLGw5dD5A

Chapter 1

History of X-Ray Crystallography

1.1 The Discovery of X-Rays

In the history of science, 1895 marks the boundary between the 19th and 20th centuries. In the small Bavarian town of Würzburg, Wilhelm Röntgen was working on the Crookes tube, one of the first electric discharge tubes. This is a vacuum tube containing two electrodes, a cathode and an anode. When a high voltage is applied between these electrodes, we see coloured rays running from the cathode to the anode, called cathode rays. This beam is an electron beam, but in 1895 the electron particle was not yet known. Like many scientists at the time, Wilhelm Röntgen wondered about the nature of these cathode rays. He tried to find out whether something was coming out of the tube, such as the cathode rays themselves or something else. To mask the light coming out of his tube, he wrapped it in black paper, switched on his tube and noticed a flashing light on a fluorescent screen that he had placed on a table a few metres away. He had just discovered what he called X-rays, X for the unknown in mathematics. It took Röntgen three days of

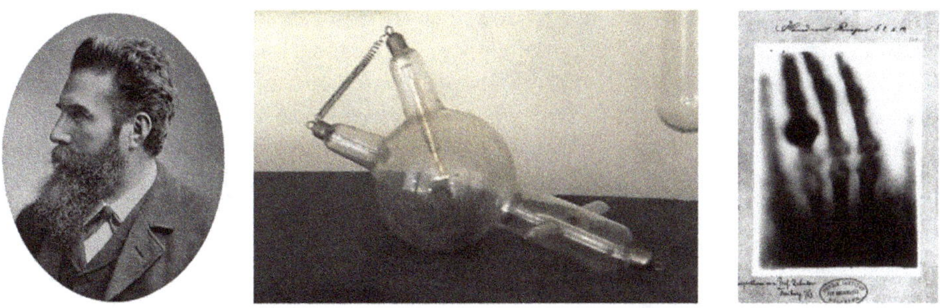

FIG. 1.1 – Left: Wilhelm Conrad Röntgen (Wikipedia: https://fr.wikipedia.org/wiki/Wilhelm_Röntgen); Center: Crookes tube (Wikimedia: https://commons.wikimedia.org/wiki/File:Crookes_tube.jpg); Right: First X-ray of the hand of Anna Bertha Ludwig Röntgen taken on December 22, 1895 (https://fr.wikipedia.org/wiki/Wilhelm_Röntgen#/media/Fichier:First_medical_X-ray_by_Wilhelm_Röntgen_of_his_wife_Anna_Bertha_Ludwig's_hand_-_18951222.gif).

DOI: 10.1051/978-2-7598-3501-0.c001

experiments in order to fully characterise these X-rays. In particular, he observed that they were very penetrating, and he took the famous radiograph of his wife Bertha's hand, now known as the first X-ray radiograph (figure 1.1).

1.2 The Nature of X-Rays

For the next seventeen years after Wilhelm Röntgen's discovery, scientists investigated the nature of X-rays. It was already known by the that:

-they are penetrating;
-they are not reflected or refracted like visible light;
-they cannot be diffracted through a small opening;
-they are not deflected by electric or magnetic fields.

So, what is the nature of these rays? Röntgen's first hypothesis, in his very first paper in 1895, was that they were longitudinal waves, like acoustic waves. Three years later, in 1898, J.J. Thomson, famous for discovering the electron, suggested that they were electromagnetic pulses. Then, in 1905, Charles Barkla, an English physicist, showed that they were transverse waves. But in 1907, William Henry Bragg, whom we will meet again later, thought that they were neutral particles. Finally, in 1912, Arnold Sommerfeld, a German theoretical physicist, hypothesised that X-rays were electromagnetic waves and deduced from his experiments that they had a wavelength of about 0.4 Å.

1.3 The Origins of Crystallography

At the end of the 18th century, René-Just Haüy postulated that a crystal is made up of small bricks, the 'integrating molecules', which periodically pack together in the three directions of space. He used drawings to describe how this packing could explain the regular shape and facets of crystals (figure 1.2). During the 19th century, the atomic theory grew, but the idea that a crystal was a regular arrangement of atoms remained 'the most likely hypothesis', according to George Wulff, a Russian crystallographer. According to this hypothesis, the integrating molecules of Haüy (now called cells) are filled with atoms or groups of atoms. But back then, it was impossible to know how these atoms were arranged or how far apart they were, because there was no possibility to measure them.

At this point, let us go back in time a little and meet William Barlow, an amateur crystallographer who, simply by thinking about the size of small spheres, was able to correctly predict very simple structures such as that of cesium chloride, CsCl, sodium chloride, NaCl, or even that of diamonds, which he correctly described in 1897 with William Pope, a professional crystallographer. In 1912, in France, Jean Perrin provided different ways of estimating Avogadro's number, the number of atoms per mole. Thus, by hypothesising the structure of crystals and determining

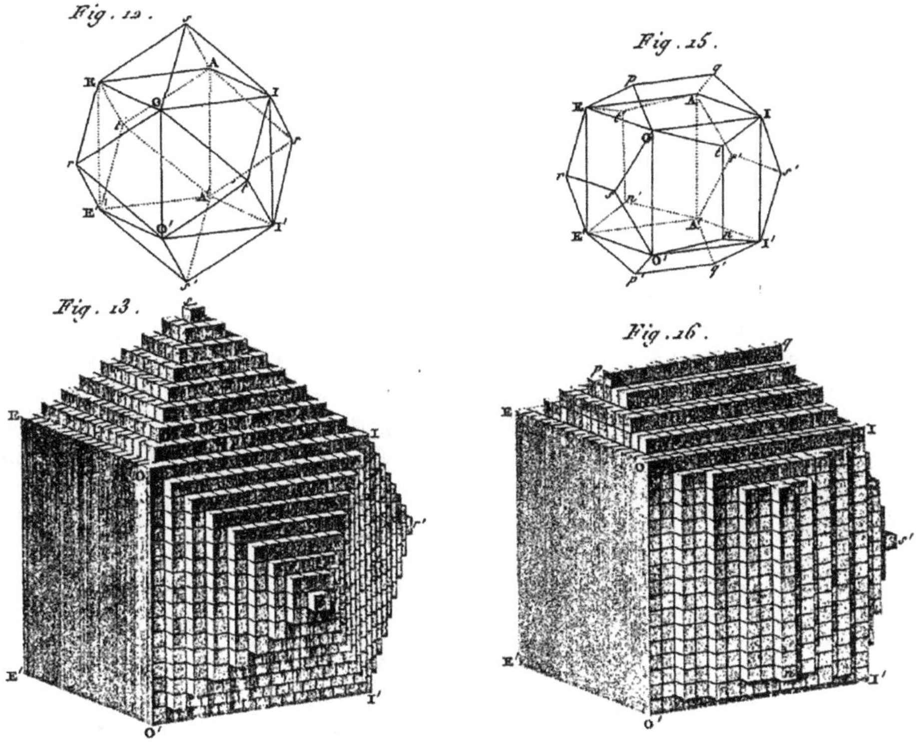

FIG. 1.2 – Molecules in a cubic crystal, 'Traité de Minéralogie' (1801), René Just Haüy (1743–1822) (https://fr.wikipedia.org/wiki/René_Just_Haüy#/media/Fichier:Hauy_Molécule_intégrante.jpg).

the volume occupied by a known number of atoms, it was possible to estimate the interatomic distances, which are of the order of ångström.

1.4 The Discovery of Diffraction

In January 1912, in the park of the Ludwig Maximilian University in Munich, Paul Ewald, a doctoral student of Arnold Sommerfeld who was working on light, and Max Laue (ennobled in 1913 as Max von Laue), a theoretician also working in Sommerfeld's team, had a famous discussion. Ewald was studying the propagation of light in what was then called a crystal, an arrangement of small, closely spaced atoms that scattered light waves. During a discussion with Ewald, Max Laue understood that crystals were like lattices with periods of a few tenths of a nanometre. Knowing the wavelengths of X-rays, he asked Ewald: "What would happen if your light had a much shorter wavelength?" Ewald replied that he didn't know what would happen. This was likely the moment when Max Laue had the idea of trying to diffract X-rays through crystals. However he encountered an obstacle: Arnold Sommerfeld, his boss, did not want Laue to carry out his diffraction experiment. He had several reasons for this, but the main argument was that he believed that thermal motion – which causes atoms to move around their equilibrium position – would cancel out the interference phenomenon that Laue expected. Sommerfeld was wrong, but that will be understood later. Fortunately for Max Laue, in April 1912 Sommerfeld was away from his laboratory and when the cat was away... the mice played Laue's diffraction experiment. The mice were Walter Friedrich, Sommerfeld's assistant, who had prepared the set-up, and Paul Knipping, a doctoral student of Röntgen's who was familiar with X-rays.

They set up an experiment using a vacuum tube, a lead plate with a small hole through which the X-rays could pass, a copper sulphate crystal, and a photographic plate. After a few unsuccessful attempts, they saw large spots on the plate: these were the first diffraction spots. After improving their setup, they obtained a very nice photograph in which we can see fine diffraction spots, later called Bragg reflections, each with its own intensity, and closely related to the symmetry of the crystal (figure 1.3).

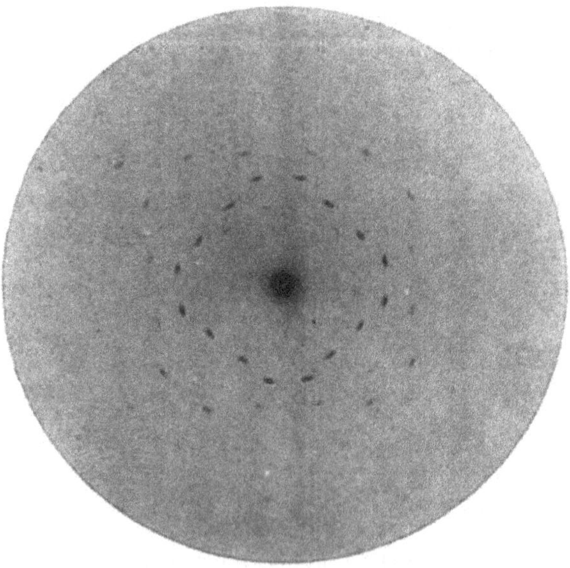

FIG. 1.3 – Laue diagram. Diffraction pattern of a blende crystal, ZnS, illuminated by an X-ray beam parallel to a fourth-order axis of rotation present in the crystal (from Friedrich *et al.*, *Annalen der Physik* 346, 971–988 (1913)).

Max Laue then thought that these diffraction spots were due to fluorescence, the emission of radiation by matter when illuminated with higher energy radiation. This explanation was incorrect and the Bragg family understood what was going on.

In 1908, Sir William Henry Bragg hypothesised that X-rays could be particles. He was wrong. In 1912, his son Lawrence repeated Laue's experiment and explained the result by describing X-rays as waves. At this point, he disagreed with his father, which made it more difficult for him to write his first paper. Lawrence Bragg's idea was that the X-rays were reflected like a mirror by the different planes of the crystal lattice and that the reflected waves interfered with each other. In November 1912, *i.e.* six months after Laue's paper, he wrote the famous formula, now known as Bragg's law:

$$2d\sin\theta = \lambda$$

where θ is the angle between the incident beam and the reticular plane, d is the distance between two planes and λ is the wavelength of the X-ray beam. This formula gives the condition for constructive interference. This results in the appearance of diffraction spots (see chapter 4) on the film, known as 'spots' or 'Bragg reflections'.

1.5 The First Structures

Lawrence Bragg understood (and here lies his genius) that crystal structures could be determined by analysing the positions and intensities of the diffraction spots. His pragmatic approach differed from that of the German school, which was more focused on the mechanism of interaction between X-rays and matter. The Braggs used the structures predicted by William Barlow and reproduced the diffraction patterns by trial-and-error methods, trying different atomic positions and changing them until the intensities matched those predicted. Using this approach, they determined the structure of sodium chloride NaCl, zinc sulphide ZnS and diamond in 1913, followed by increasingly complex structures, such as that of fluorite CaF2 in 1914. Other researchers began to use this method, such as the Norwegian Vegard, who determined the structure of titanium oxide in 1916 during the First World War (figure 1.4).

These structures now seem extremely simple, perhaps a little outdated, but for their time they were revolutionary and challenged certain preconceptions. In 1927, a chemist wrote: "Some books are lies through and through. Professor Bragg states that there is not one molecule of NaCl in sodium chloride, a statement that is totally offensive to common sense, it is absurd to the last degree". Some researchers were still convinced that there were stacks of NaCl molecules in sodium chloride, whereas Bragg's structure showed that the sodium atoms and chlorine atoms were regularly packed together and that there were therefore no NaCl molecules in the solid. In 1920, 50 structures were determined using the Bragg method, and in 1925, 600, including 150 organic molecules.

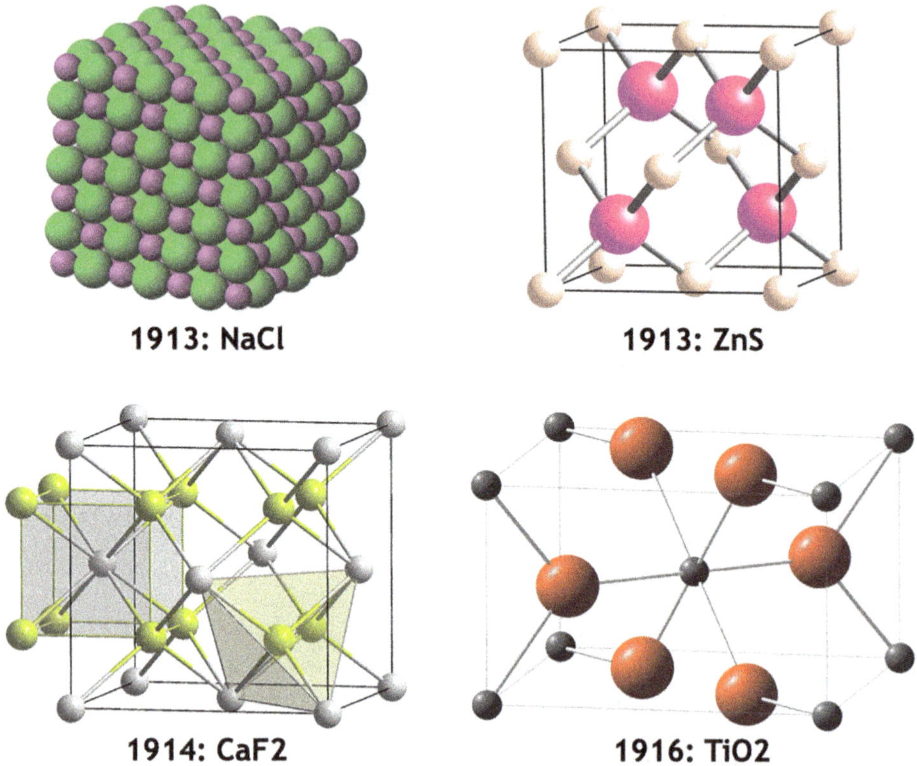

FIG. 1.4 – First crystal structures determined by X-ray diffraction (NaCl: https://commons.wikimedia.org/wiki/File:NaCl.png; ZnS: https://fr.m.wikipedia.org/wiki/Fichier:Boron-arsenide-unit-cell-1963-CM-3D-balls.png; CaF2: https://fr.m.wikipedia.org/wiki/Fichier:CaF2_polyhedra.png; TiO2: https://commons.wikimedia.org/wiki/File:Titaniumdioxide_unit_cell_rutile.svg).

1.6 Sir William Henry Bragg's Brilliant Move

The trial-and-error approach used for the first molecular structures could not be used for more complex structures, so another way of solving these structures had to be found. Since 1912 and the discoveries of his son Lawrence, Sir William Henry Bragg had realised his mistake and agreed with his son that X-rays were short wavelength electromagnetic waves. In 1915, he wrote a famous article linking the mathematics of the Frenchman Joseph Fourier to diffraction. He had just discovered that X-ray diffraction could be described by a Fourier transform of the electron density of a crystal (see chapter 8). In brief, figure 1.5 shows small clusters of electron densities arranged periodically as in a crystal. This arrangement allows a Fourier series expansion whose amplitude of the first, second, third and so on order terms is shown in the diagram (figure 1.5). Sir William Henry Bragg showed that these terms are related to the intensity of the Bragg spots visible in the diffraction

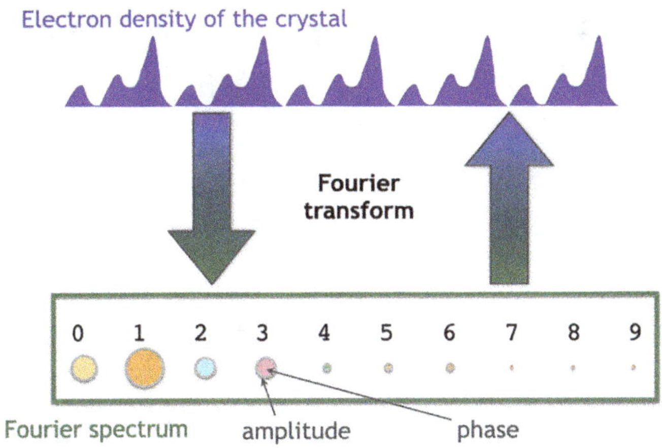

FIG. 1.5 – Sir William Henry Bragg's use of the Fourier transform. In this representation of the Fourier spectrum, the diameter of the circle represents the amplitude, and the colour symbolically represents the phase of the waves in the series (see chapter 8 for more details).

pattern. This made structure determination easier in theory, but still difficult in practice because Fourier series calculations are extremely time-consuming to do 'by hand'. It was not until the development of the first computers, in the 1970s, that this technique came into its own.

Around the years 1925–1930, the first results were structure determinations that can be called *ab initio*, because they do not require any *a priori* knowledge of the structure, unlike the first structures of W. Barlow. In 1929, W. Zachariasen solved the structure of potassium chlorate, and in 1931 Kathleen Lonsdale solved the structure of benzene hexachloride, confirming the cyclic nature of the molecule. In 1935, phthalocyanine was solved, a molecule consisting of about 60 atoms, but flat and symmetrical. Then, in 1945, Dorothy Hodgkin determined the structure of the cholesterol molecule and its 74 atoms, followed in 1954 by the structure of vitamin B12, which has about two hundred atoms (figure 1.6).

1.7 Biology Comes into Play

Crystallography is based on the production of crystals of the target under investigation. To study the structure of a macromolecule by X-ray crystallography, the first bottleneck is crystallisation. Since biological macromolecules are large, often flexible objects, this step can be quite difficult (see chapter 5). In the case of proteins, the very first observation was made in 1840 by Friedrich Ludwig Hünefeld, who accidentally discovered the formation of crystalline material in the form of small plate-like crystals in dried blood samples (figure 1.7, left). This observation

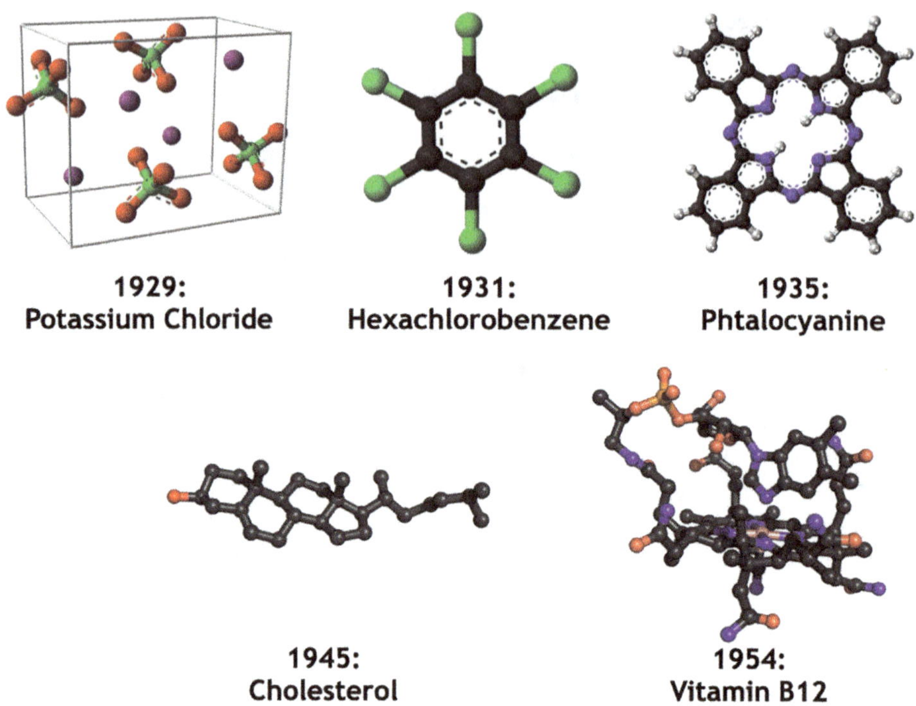

FIG. 1.6 – Evolution of the complexity of molecular structures solved by X-ray crystallography (potassium chloride: https://commons.wikimedia.org/wiki/File:Potassium-perchlorate-unit-cell-3D-balls-perspective.png; hexachlorobenzene: https://fr.wikipedia.org/wiki/Fichier:Hexachlorobenzene-3D-balls-B.png; phtalocyanine: https://commons.wikimedia.org/wiki/File:Phthalocyanine-3D-balls.png; cholesterol: https://wa.wikipedia.org/wiki/Im%C3%A5dje:Cholesterol_molecule_ball.png; vitamin B12: https://commons.wikimedia.org/wiki/File:Cyanocobalamin-3D-sticks.png).

demonstrated that protein crystals could be obtained by controlled evaporation, *i.e.* the slow dehydration of a concentrated solution, which remains the basis of most current crystallisation techniques (see chapter 5). In the following years, haemoglobin crystals were observed in many animal species, the characteristics of which were summarised in a book entitled 'Die Blutkrystalle' (Blood Crystals), published in 1871 by William T. Preyer, a professor at the University of Jena. In 1855, Theodor Hartig discovered a second family of proteins in crystalline form, the reserve proteins of the Brazil nut Bertholletia excelsa (figure 1.7, right). Crystals have also been observed for equivalent proteins in related plant species.

In the first half of the 20th century, many proteins were crystallised (trypsin, carboxypeptidase, catalase, chymotrypsin, ribonuclease, urease, ...). The main process for obtaining crystals, described by Thomas Osborne in 1907, is based on protein extraction from hot saline solutions (40–60 °C) followed by slow cooling

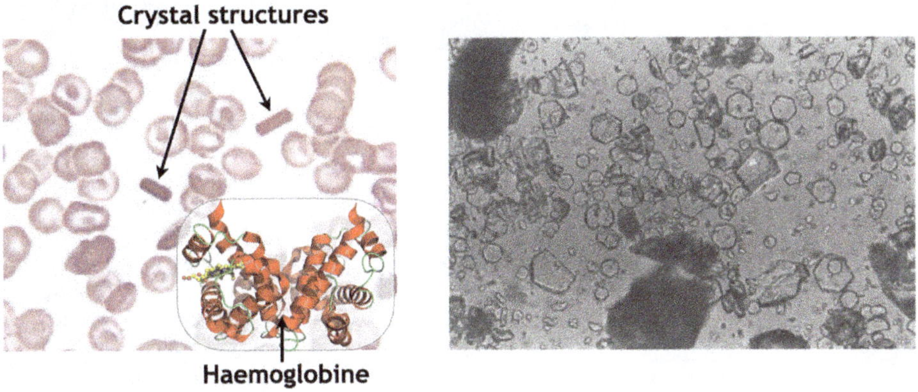

FIG. 1.7 – Microscopic image, left, of haemoglobin crystals (red platelets) among red blood cells (https://imagebank.hematology.org/image/3787/hemoglobin-c-crystals–2?type=upload), with the haemoglobin molecule solved by Max Perutz as an inset (file 2dhb.pdb); on the right, crystals of Brazil nut storage proteins, known as excelsin (after Mc Pherson, 1991).

down to room temperature. Many researchers have emphasised the importance of temperature, pH, ionic strength, organic solvents and phase changes in crystallisation. These principles still prevail in modern crystallisation experiments. At that time, crystallisation was of interest to scientists as a method of purifying a sample and ensuring the homogeneity of preparations. The crystallisation of enzymes such as urease by James B. Sumner in 1926, pepsin in 1930, trypsin in 1931 and chymotrypsin in 1933 by John H. Northrop, earned them the Nobel Prize in Chemistry in 1946. The combination of crystallisation and biochemical studies enabled them to demonstrate that a perfectly pure enzyme was active and that the catalytic entity was the protein.

A breakthrough in the study of protein crystals came in 1934 when John Desmond Bernal and his student Dorothy Hodgkin performed the first diffraction experiment using protein crystals surrounded by their mother liquor in a capillary tube. Previously, the protein crystals had been exposed to air before being exposed to X-rays, which dehydrated them and caused no observable diffraction. They found that the diffraction patterns were much better than with dried crystals. This was the first X-ray diffraction pattern of a protein crystal. It was also the beginning of biological crystallography and probably the birth of structural biology. However, analysing the diffraction pattern of a protein crystal was a complex problem to solve. The large number of diffraction spots (reflections) required the development of suitable data acquisition methods and cameras, such as the Weissenberg chamber, developed in 1924, which was able to record a large number of reflections simultaneously. The big question, however, was how to determine the phase associated with each reflection.

X-ray crystallography and its succession of Nobel prizes (https://www.nobelprize.org/)			
Year	Nobel price	Authors	Reason
1901	Physics	Wilhelm Conrad Röntgen	Discovery of X-rays
1914	Physics	Max von Laue	Discovery of X-ray diffraction by crystals
1915	Physics	Sir William Henry Bragg, William Lawrence Bragg	Contribution to crystal structure analysis using X-rays
1939	Physics	Ernest Orlando Lawrence	Invention and development of the cyclotron, and the results obtained with it, especially in connection with radioactive elements
1946	Chemistry	James Batcheller Sumner	Discovery that enzymes can be crystallised
		John Howard Northrop, Wendell Meredith Stanley	Preparations of enzymes and viruses in pure form (by crystallisation)
1954	Chemistry	Linus Carl Pauling	Nature of the chemical bonding and elucidation of the structure of complex substances
1962	Medicine	Francis Harry Compton Crick, James Dewey Watson, Maurice Hugh Frederick Wilkins	Discoveries about the molecular structure of nucleic acids and their importance for the transfer of information in living material
1962	Chemistry	Max Ferdinand Perutz, John Cowdery Kendrew	Study of the structures of globular proteins
1964	Chemistry	Dorothy Crowfoot-Hodgkin	Structure determination of relevant chemicals using X-ray techniques
1982	Chemistry	Aaron Klug	Development of crystallographic electron microscopy and structure elucidation of biologically relevant protein/nucleic acid complexes
1985	Chemistry	Herbert Aaron Hauptman, Jerome Karle	Outstanding results in the development of direct methods for crystal structure determination
1988	Chemistry	Johann Deisenhofer, Robert Huber, Hartmut Michel	Determination of the three-dimensional structure of the photosynthetic reaction centre
1997	Chemistry	Paul Delos Boyer, John Ernest Walker	Elucidation of the enzymatic mechanism of ATP synthesis
		Jens Christian Skou	First discovery of an ion-transporting enzyme, the Na^+, K^+-ATPase
2003	Chemistry	Peter Agre, Roderick MacKinnon	Determination of the structure of ion channels in cell membranes
2009	Chemistry	Venkatraman Ramakrishnan, Thomas Arthur Steiz, Ada Yonath	Studies of the structure and function of the ribosome

From 1937, Max Perutz worked on the analysis of protein crystals and obtained diffraction images of haemoglobin crystals. In 1945, John Kendrew began working with Perutz, working on myoglobin crystals, a protein that has the advantage of being four times smaller than haemoglobin. In 1953, a new shift in biological crystallography occurred when Perutz used the isomorphic replacement method to solve the phase problem of haemoglobin diffraction spots. He successfully incorporated benzoylmercury into haemoglobin crystals, recorded the diffraction patterns and determined the position of the heavy atom, mercury, from the difference between the native and the derived data (see §11.1). In 1957, Kendrew solved the structure of myoglobin using this approach and achieved atomic resolution in 1959, the same year as Perutz solved the structure of haemoglobin. In 1962, they shared the Nobel Prize in Chemistry for these studies on globular protein structures. The case of the structure of DNA is somewhat different since the data used were diffraction data from B-form DNA fibres (see §1.8).

Although the structure of several proteins was solved after that of haemoglobin and myoglobin, X-ray analysis of protein crystals presented major problems, both in the preparation of heavy-atom derivatives and in the need to use several derivatives to overcome phase ambiguity (see §11.1). To limit the number of heavy atom derivatives, Johannes Bijvoet proposed in 1954 to exploit the anomalous scattering signal by using a wavelength similar to that of the absorption edge of the heavy atom present in the crystal (see §11.2). This approach was successfully applied first to an organic molecule by Gopalasamudram Ramachandran in 1956 and then to a protein by David Blow and Michael Rossman in 1959. In 1962, David Blow and Michael Rossman proposed an approach based on prior knowledge of a fragment of the molecule contained in the crystal or a similar molecule, using the Patterson function and non-crystallographic symmetries. This method was called molecular replacement by Michael Rossman in 1972 (see chapter 10). Whatever the method, the computations required to solve a structure were limited by the low power of the computers of the time, and the sharing of resources made it possible to speed up and optimise the methods of solving structures (see §1.9 for the role of CCP4). Biological crystallography had only to develop and focus on increasingly complex proteins and molecular assemblies: membrane proteins, nucleoprotein assemblies, viruses... (see box on **X-ray crystallography and its succession of Nobel prizes**).

1.8 Rosalind Franklin and the Mystery of Cliché 51

In the 1950s, researchers who were not yet called biocrystallographers became interested in other types of molecules, namely nucleic acids. Since the work of Oswald Avery in 1944, it has become increasingly clear that deoxyribonucleic acid, or DNA, was the carrier of heredity. However, the structure of this complex molecule remained unknown, even though in 1937 Florence Bell and William Astbury at the Textile Physics Laboratory in Leeds had used X-rays to suggest that the bases of DNA were stacked like coins – a pile of pennies – 3.34 Å apart.

Work on DNA structure only resumed shortly after the war at King's College London, then headed by John Randall, who commissioned Maurice Wilkins and his student Raymond Gosling to study the structure of DNA. In 1950, they obtained the first photograph of crystallised DNA from calf thymus, in its A-form (figure 1.8): a beautiful result. Randall then invited a young Englishwoman, Rosalind Franklin, who had just returned from a post-doc in Paris, to King's College. He asked her to work with Gosling on the DNA without telling Wilkins about it. This management error led to Wilkins and Franklin becoming rivals in the same laboratory.

This situation was further exacerbated by the involvement of two researchers inspired by Wilkins' results, Francis Crick and James Watson. They started working on DNA at the famous Cavendish Laboratory in Cambridge, then headed by Lawrence Bragg. On the other side of the Atlantic, the great Caltech chemist Linus Pauling, who had discovered the α-helix of proteins in 1951, was also looking for a model for the DNA molecule. The ingredients were there for a race to find a model.

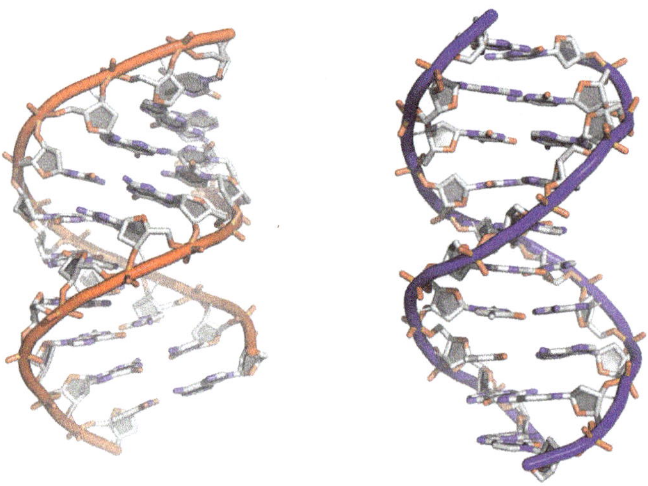

FIG. 1.8 – A-form (red, file 1d13.pdb) and B-form (blue, file 5dnb.pdb) of DNA.

However, it was the researchers at King's College who carried out the diffraction experiments and Rosalind Franklin who made the most important advances. She was the first who identify another form of DNA, called the B-form, which was observed at higher humidity than the A-form. It appeared as a disordered phase with no Bragg spots (figure 1.9). From her diffraction patterns, she realised that this form of DNA was helical and from her experiments, she correctly deduced that the hydrophilic phosphate groups were located on the outside of the molecule.

The famous plate 51 (figure 1.9) was taken by Raymond Gosling and Rosalind Franklin on the 1st of May 1952, after a 60-h exposure. Before leaving for another laboratory, she sent it to Wilkins at the end of January 1953, and he showed it to Watson during a visit to King's on 30 January, without Rosalind's knowledge. This is one of the turning points in the story described by James Watson in his book 'The Double Helix' (Watson, 1968): "As soon as I saw this image, I was speechless and my pulse quickened [...] The black cross of reflections that dominated the image could only have come from a helical structure." This cross is indeed characteristic of diffraction by a helix, whose parameters can then be found: a step of 3.4 nm and a diameter of 2 nm. Thanks to this and other information, also from King's College, Watson and Crick were certain that DNA had a helical shape, and a month later, on 28 February, after many tests, they found that it was the pairing, and not the stacking, of bases that allowed them to understand the structure of DNA.

When Watson and Crick received their Nobel Prize in 1962, they made no mention of Rosalind Franklin's work. Only Wilkins mentioned her briefly, reporting her death in 1958. She had died of ovarian cancer, possibly as a result of her X-ray experiments. A controversy over her role in the discovery arose when James Watson, in his 1968 book, painted an unfavourable, even libellous, picture of Rosalind Franklin, downplaying her work and abilities. Colleagues and friends came to her defence, such as Aaron Klug (Klug, 1968), a future Nobel Laureate in Chemistry, and Anne Sayre, who wrote a book about her in 1975 (Sayre, 1975). Since then, several books have fully rehabilitated Rosalind Franklin (Williams, 2019; Maddox, 2003), showing the importance of her work in the discovery process, and Watson and Crick have acknowledged that they discredited her.

From a crystallographer's point of view, it is interesting to note that this discovery was not made by 'classical' analysis of DNA crystals (the A-form), as Rosalind Franklin attempted, but by working on a disordered phase (the B-form) and testing models, as Watson and Crick did. Since the DNA backbone was ultimately quite simple (a twisted ladder), this was the method that led to the solution. It was only about ten years later that the atomic structure was solved using the methods explained in this book.

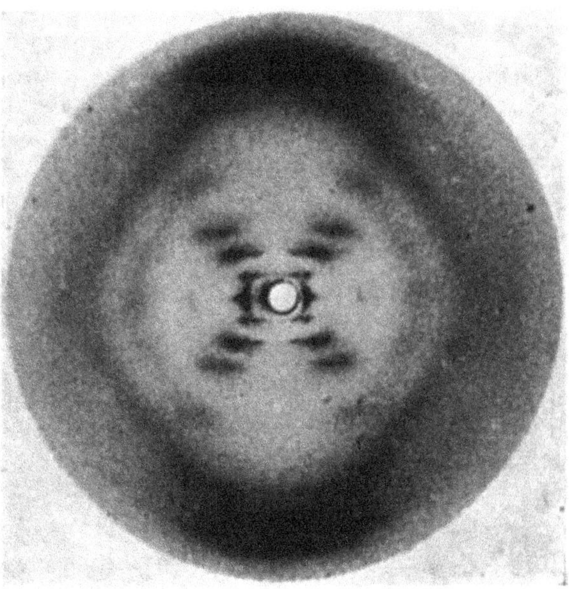

FIG. 1.9 – Plate 51 of the B-form of DNA (https://en.wikipedia.org/wiki/Photo_51#/media/File:Photo_51_x-ray_diffraction_image.jpg).

1.9 CCP4: Collaborative Computational Project No. 4

To conclude this journey through history, let's go back to the 1970s. Biological crystallography was born. Living molecules are very different from those studied in chemistry: they are huge and have very specific geometric and qualitative properties. Crystallographers had to rethink the method to adapt it to macromolecules; they had to rethink the solutions that would allow them to solve the three-dimensional structures. Computers exist, but biological crystallography requires considerable computing. The UK is ahead of the game, and the UK Science Research Council recognises the need to build collaborations to optimise the use of computational tools. It funds a series of projects called CCPs, which stand for 'Collaborative Computational Project'. The one that interests us and has played a major role in biological crystallography is project number 4: CCP4, which remains a reference in biological crystallography (http://www.ccp4.ac.uk/).

The CCP4 project began in 1979. The interests associated with this project were primarily those of the small community of macromolecular crystallographers, mainly based in Cambridge and Oxford. These crystallographers needed help in developing new algorithms, but also in installing and maintaining the software. It was the 1970s, and there was a sense of community in the air. So the partners in the CCP4 project decided that there would be no single group with a single owner of good ideas and that there would be no question of intellectual property. The Fortran language made it easy for developers to hack each other. In an era before the internet, the funds allocated to CCP4 were mainly used for face-to-face meetings. The spirit was, and still is, of mutual support, and also the so-called 'bottom-up', where users express their needs and developers look for ways to meet them.

The objectives of CCP4 are:

- to promote the collaborative development of software for macromolecular crystallography;
- to provide software for macromolecular crystallography steps;
- to promote the teaching of macromolecular crystallography.

This project has supported collaborations between researchers working on software suitable for protein crystallography, initially in the UK. The aim was to build up a comprehensive collection of software to meet the needs of UK groups, and then rapidly to meet the needs of groups around the world.

At each stage of crystallographic structure resolution, it is necessary to refer to the properties of:

- diffraction data;
- scattering data of the atoms;
- the crystals: such as their symmetries;
- the Fourier transforms.

CCP4's answer was the introduction or generalisation of:

- common libraries that can be used by all software;

- standard formats for the reflections, first the LCF format and then the current MTZ format. In short, these are binary files which can be read directly by software and which contain all the information about the diffraction data;
- a suite of modular programs: any new software can be integrated without modifying the rest;
- a discussion forum: in particular the ccp4bb 'mailing list', for the ccp4 bulletin board, allows questions to be put to the whole community. This leads to interesting ideas and sometimes exciting expert discussions;
- a training weekend: each year a theme is explored in depth, allowing users to meet the authors of the software and learn more about the underlying theory.

Since 1979, CCP4 has gradually become an international reference in crystallography. This type of collaborative organisation has proved effective in other cases such as the UNIX operating system, the development of the Blender 3D software, or the Wikipedia resource.

In summary, the impact of CCP4 can be summarised as follows:

- acceleration of software development;
- acceleration of methodology development;
- stable international standards;
- sharing of data, knowledge and maintenance;
- a sense of mutual support.

An important consequence is also the existence of a real international community that can mobilise around needs such as the development of synchrotron radiation centres, or around new methodological issues, such as those currently raised by electron cryo-microscopy.

Finally, the timeline below illustrates the main stages in the development of the method (in yellow), structural analysis (in blue) and biological crystallography itself (in green).

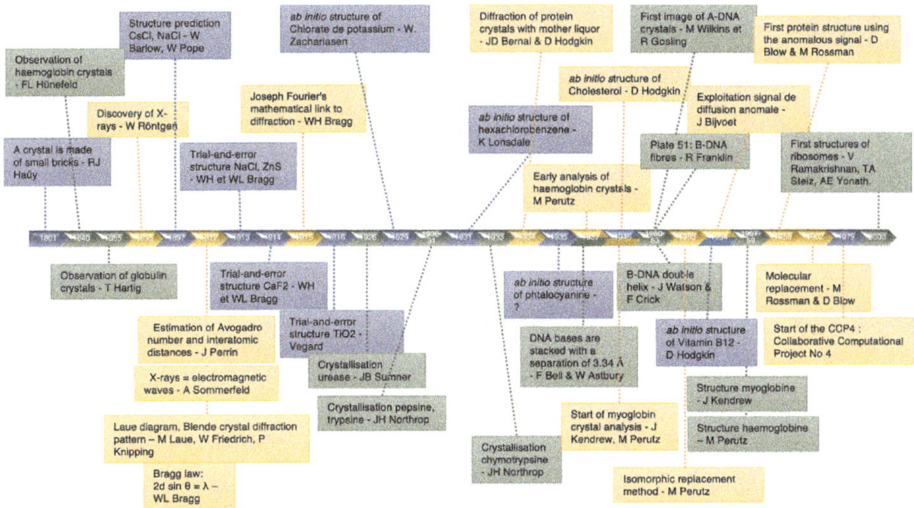

References

Publications

Brooks-Bartlett J.C., Garman E.F. (2015) The nobel science: One hundred years of crystallography, *Interdiscip. Sci. Rev.* **40**, 244–264.

Giege R. (2013) A historical perspective on protein crystallization from 1840 to the present day, *FEBS J.* **280**, 6456–6497.

Klug A. (1968) Rosalind Franklin and the discovery of the structure of DNA, *Nature* **219**, 808–810.

Maddox B. (2003) *Rosalind Franklin, the dark lady of DNA*. Harper Collins Publishers.

Mc Pherson A. (1991) A brief history of protein crystal growth, *J. Cryst. Growth* **110**, 1–10.

Sayre A. (1975) *Rosalind Franklin and DNA*. W.W. Norton, New York.

Watson J.D. (1968) *The double Helix*. Simon and Schuster, New York.

Williams G. (2019) *Unravelling the double helix*. Pegasus books Ltd, New York.

Related videos

History of X-ray crystallography: the early days (Histoire de la cristallographie aux rayons X : les débuts)
https://youtu.be/uBMipJbzz48

History of X-ray crystallography: continued (Histoire de la cristallographie aux rayons X : suite)
https://youtu.be/Xzgv5wtgcco

The role of the CCP4 (Le rôle du CCP4)
https://youtu.be/unlDN2HfLkY

Chapter 2

Sample Preparation

2.1 Knowing Your Sample

To crystallise a biological macromolecule and carry out a structural study, the amount of sample required is typically in the order of a few milligrams. It can take several weeks or months to prepare such a large amount. In the case of a protein, a tricky first step is to decide which part of the sequence to clone, produce and purify in order to maximise the likelihood of obtaining crystals. Protein production, purification and crystallisation require a number of conditions (figure 2.1):

- the production process requires the knowledge of the sequence of the gene encoding the protein of interest. Knowledge of the post-translational modifications is also crucial for the proper choice of the production host organism;
- the purification step relies on the solubility and stability of the selected part of the protein, which determines the amount of the target protein that can be obtained;
- the crystallisation step is only possible if the chemical purity, the conformational homogeneity and the solubility of the sample are suitable.

Step	Prerequisites
Production	Sequence, Post-translational modifications
Purification	Solubility, Stability
Crystallisation	Purity, Solubility

FIG. 2.1 – Steps in preparing a biological sample for crystallisation.

2.1.1 Prediction of Structured Regions of a Protein

The sequence of the gene encoding the protein of interest, either the nucleotide sequence of the gene or the amino acid sequence of the protein, is the most important information to know. In some cases, it may be useful to adapt the nucleotide sequence so that it can be produced by simple microorganisms such as

DOI: 10.1051/978-2-7598-3501-0.c002

bacteria or yeast. For example, in the case of a human protein, the nucleotide sequence may contain rare codons that are not present in most common bacteria used for the production step. As a result, the translation step of the protein by the ribosome is significantly slowed down and the amount of protein produced is reduced. To overcome this problem, the gene must be modified by replacing the rare codons with codons that are more common in the bacteria.

In order to anticipate the optimal crystallisation conditions, it is important to have an insight into the structured and unfolded regions and then attempt to predict these regions from the individual protein sequence. Amino acids are the basic building blocks of the protein sequence. They have an NH2 amine functional group, a COOH carboxylic acid group, and an R-side chain (see box on the **20 natural amino acids**). These side chains have different chemical properties. There are different classification criteria, but we find side chains that are:

- polar, positively charged (Arg, His, Lys) or negatively charged (Asp, Glu);
- polar but not charged (Asn, Gln, Ser, Thr);
- non-polar and hydrophobic (aliphatic: Ala, Val, Leu, Ile, Met; aromatic: Phe, Tyr, Trp);
- with specific properties, they are then classified separately (Gly, Pro, Cys).

The **20 natural amino acids** found in proteins are in alphabetical order. In the table below, the atoms are coloured yellow for carbon, blue for nitrogen, red for oxygen, orange for sulphur and white for hydrogen. Each amino acid is oriented from left to right, N-terminal to C-terminal, with the nitrogen at the bottom left, the carbonyl at the bottom right, and the side chain at the top.

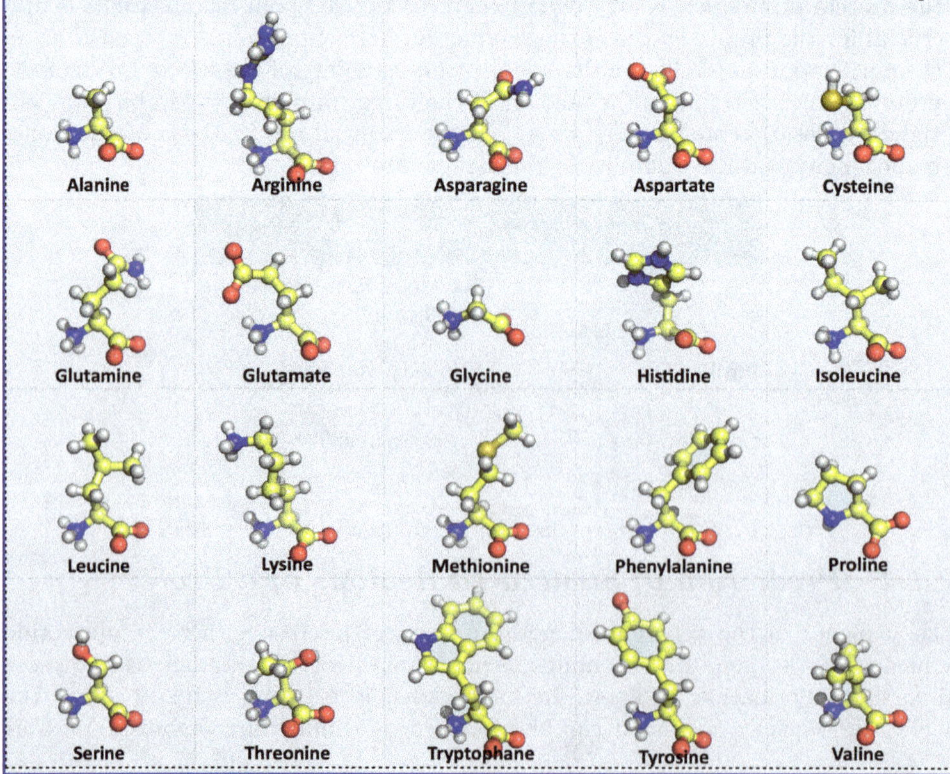

Depending on their side chain, amino acids have different chemical (or electrochemical) properties, which means different structural properties for the protein. Based on already-known protein structures, it is possible to classify different structural properties according to the side chains of the amino acids. This is called a secondary structure. For example, the tendency of an amino acid to be buried or on the surface, or to be found in one type of structure rather than another, is linked to the properties of its side chain. Different types of secondary structures have been described, such as α-helices, β-strands and various types of turns (figure 2.2).

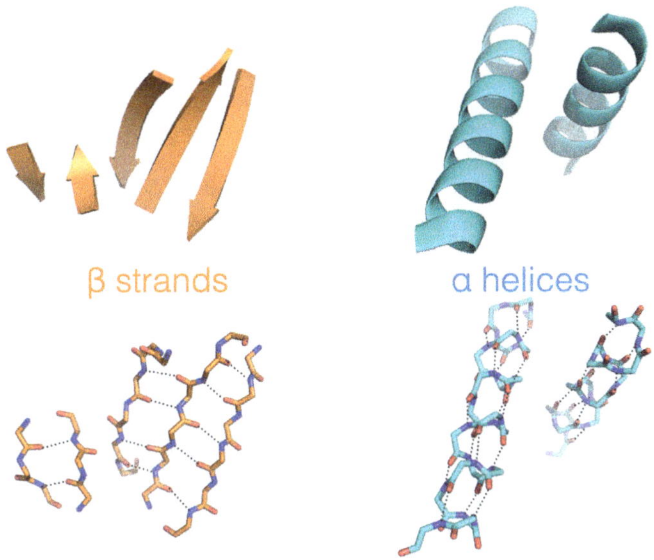

FIG. 2.2 – Typical secondary structures, shown above in cartoon representation and below in stick representation of the backbone.

However, the prediction of secondary structures from the amino acid sequence alone is rather unreliable. In addition to the observed tendency of an amino acid to occur in certain secondary structures, the order of the amino acids in the sequence itself is an important piece of information. The already sequenced genomes provide a database that allows the sequence of the protein under investigation to be compared with that of other organisms. Sequences that are similar to the protein under study and found in other organisms indicate that their three-dimensional structures are similar. To carry out sequence alignments, there are software programs that search databases for similarities or homologies. For example, in the case of the standard BLAST (Basic Local Alignment Search Tool) software, the user provides the sequence of the protein under study. The program searches for analogous protein sequences and displays their alignment with the target sequence as a result (figure 2.3).

There are many algorithms available for predicting secondary structure based on the properties of amino acid side chains and their conservation across species. The higher the number of homologs for a given protein, the better the secondary

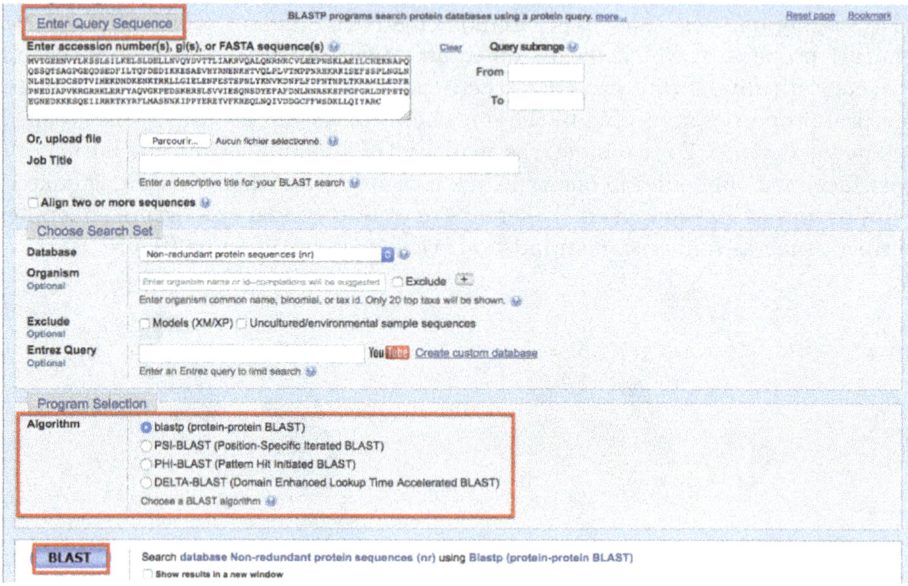

FIG. 2.3 – BLAST homologous sequence search software interface (http://blast.ncbi.nlm.nih.gov/Blast.cgi).

structure can be predicted. If there is a sufficient number of homologous sequences to the protein of interest, it is possible to define highly conserved, moderately conserved or poorly conserved regions. It is also possible to predict the presence of secondary structures such as beta strands or alpha helices. Regions rich in secondary structures generally correspond to compact regions, and those without secondary structures correspond to more disordered and flexible regions. A compact or globular domain is generally easier to produce in large quantities because of the stabilising effect of sequence folding. Conversely, unstructured regions usually do not crystallise, because the solution contains a variety of conformations and therefore lacks the homogeneity required for crystallisation.

In addition to these secondary structure predictions, newer software based on artificial intelligence and using evolutionary information can predict the 3D structure in a highly reliable way. Depending on the reliability score obtained for the 3D model, the relevance of the crystallographic study itself may be questioned.

2.1.2 Biochemical Approach: Limited Proteolysis

The principle of limited proteolysis is to incubate the target protein with a low concentration of specific enzymes called proteases (see box on the main **Proteases and limited proteolysis**). Each protease recognises a sequence of a few amino acids and cleaves the main chain at specific sites in the protein. These are typically accessible regions such as loops or other flexible regions that are not protected within folded domains. The main limitation in performing controlled proteolysis is that a small amount of the whole protein must first be produced and purified. After proteolysis, a purification step, such as gel filtration, can be used to separate the folded domains. Sequence analysis of these protected domains allows us to determine their boundaries. At this stage, one can either work directly with the separated domains obtained after proteolytic digestion or order the gene corresponding to the single domain of interest and proceed to cloning and production (figure 2.4).

Proteases and limited proteolysis

Proteases are enzymes that hydrolyse the peptide bond between two amino acids in a protein. The specificity of a protease is defined by the sequence of residues of the substrate protein located at the cleavage site. This sequence is noted from P4 to P′2, with the cleavage site located between P1 and P′1. The table below shows some of the proteases that are routinely used for limited proteolysis, together with their specificity. As specificities are not absolute, more than one residue may be indicated for a given position. Other residues may block the action of the enzyme and are crossed out at the appropriate position (https://web.expasy.org/peptide_cutter/peptidecutter_enzymes.html).

Nom	P4	P3	P2	P1	P'1	P'2
Proteinase C-Arg	-	-	-	R	-	-
Proteinase K	-	-	-	A/E/F/I/L / T/V/W/Y		
Endopeptidase N-Asp	-	-	-	-	D	-
Chymotrypsine High affinity	-	-	-	F/Y	~~P~~	
	-	-	-	W	~~M~~ ou ~~P~~	
Chymotrypsine Low affinity	-	-	-	F/L/Y	~~P~~	
	-	-	-	W	~~M~~ ou ~~P~~	
	-	-	-	M	~~P~~ ou ~~Y~~	
	-	-	-	H	~~D~~. ~~M~~. ~~P~~ ou ~~W~~	
Thermolysine	-	-	-	~~D~~ ou ~~E~~	A/F/I/L/ M/V	
Thrombine	A/F/G/I/ L/T/V/M	A/F/G/I/ L/T/V/W	P	R	~~D~~ ou ~~E~~	~~D~~ ou ~~E~~
Trypsine	-	-	-	K/R	~~P~~	
			W	K	P	
			M	R	P	
Pepsine (pH 1.3)	-	~~H~~. ~~K~~ ou ~~R~~	~~P~~	~~R~~	F/L	~~P~~
	-	~~H~~. ~~K~~ ou ~~R~~	~~P~~	F/L	-	~~Pro~~
Pepsine (pH > 2)		~~H~~. ~~K~~ ou ~~R~~	~~P~~	~~R~~	F/L/W/Y	~~P~~
		~~H~~. ~~K~~ ou ~~R~~	~~P~~	F/L/W/Y	-	~~P~~
Enterokinase	D/E	D/E	D/E	K	-	-

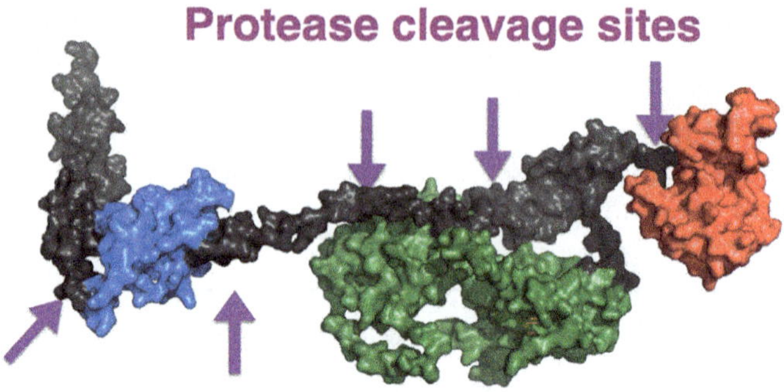

FIG. 2.4 – Representation of the globular domains of a protein (blue, green, red) preserved after digestion of the unprotected regions (grey) by a protease.

2.2 Cloning, Production, Sample Purification

The portion of the protein that can be studied by crystallography has been defined by sequence analysis. Structural studies are demanding in terms of the amount of material required. The purity of the sample is essential to obtain crystals. Milligrams or tens of milligrams of the protein must be prepared with a purity of more than 95%. Several hundred micrograms of purified protein are required to obtain a sufficient amount of sample at a concentration suitable for preliminary crystallisation experiments and to reproduce and optimise the initial leads. It is very unusual for a biological molecule to be sufficiently abundant and homogeneous in its natural state to meet the requirements of crystallisation. Alternative methods are therefore used to produce the protein of interest. The preparation of the sample can be described in three steps: cloning, production, and purification.

2.2.1 Cloning

The most common approach to producing a macromolecule is to use the natural machinery found in microorganisms such as *Escherichia coli* bacteria, yeast, insect or mammalian cells. These types of cells are called host cells and contain their own DNA. The aim is for the host cell to be able to produce the protein under study in very large quantities. This production is called overexpression, and the overexpressed protein is the target. The gene encoding the target protein is inserted into a DNA fragment, which may also contain additional fragments upstream or downstream of the target gene. For example, a fragment expressing an addressing sequence can be added to direct the target protein to a specific region of the cell. It is also common to insert a sequence encoding a 'tag'. This tag can be a short amino acid sequence, usually a histidine repeat, or a protein (MBP, GST, TRX, etc.) whose properties facilitate both the production and the purification of the target protein. The presence of the tag also facilitates purification by affinity chromatography. A cleavage site, recognised by a specific protease, allows the tag to be removed at the end of the sample preparation (figure 2.5).

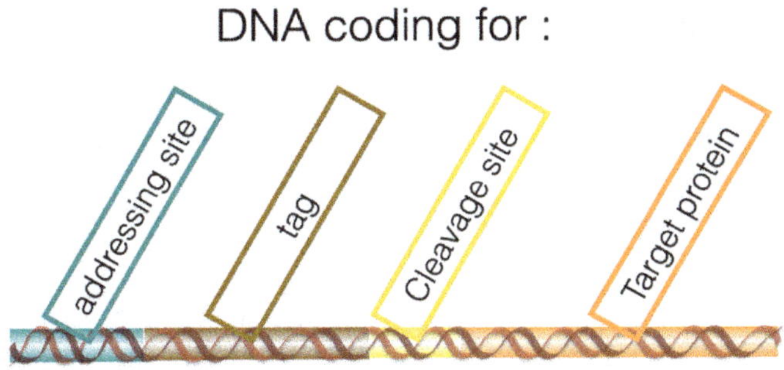

FIG. 2.5 – DNA fragment containing the target protein gene preceded by an addressing site, a tag and a cleavage site.

To enable the host cell to overexpress the target protein, the DNA fragment carrying the target protein gene is inserted into a circular piece of DNA called a plasmid during the cloning step. After cloning, the transformation, which is often considered part of cloning, consists of infecting the host cell with the plasmid carrying the target gene. After this step, the host cell can produce the target gene using its own production machinery (figure 2.6).

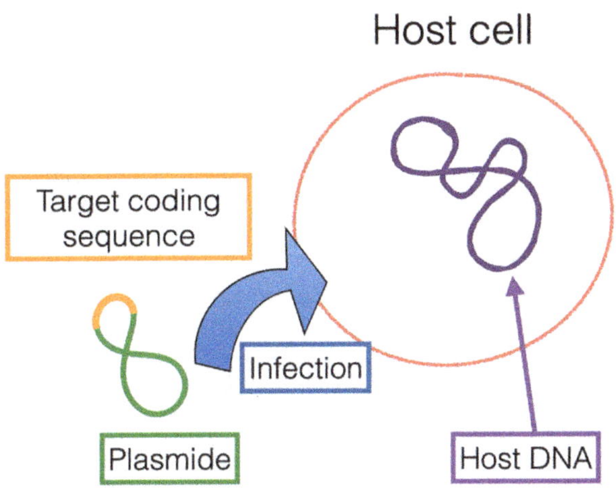

FIG. 2.6 – Transformation of the host cell with the plasmid carrying the gene sequence encoding for the target protein.

2.2.2 *(Over)production*

The host cell is transferred to a flask containing a nutrient medium suitable for its growth, *i.e.* the culture phase. During the first stage, the cells divide and proliferate rapidly, but the promoter that enables the expression of the target protein is not activated, so the target protein is not produced. When a critical mass of cells is reached, a small molecule that activates the promoter is added to the flask, inducing the host cells to produce the target. From this point on, the cells continue to divide as their transcription and translation machinery produce the target protein. After a few hours, cell division slows down due to the limited amount of nutrients in the culture medium. At this point, it is necessary to stop the production before it declines (figure 2.7).

At the end of production, the amount of sample in 1 L of culture varies from a few milligrams to a few hundred milligrams. This depends on the target protein itself and the optimisation of the protocol. In any case, the amount of the target molecule is very small compared to all the material present in the culture medium: cell walls, DNA, proteins, sugars, host lipids, etc. This means that the purification of the protein is necessary.

In some cases, the production step must be adapted to incorporate specific amino acids into the protein of interest, such as selenomethionines (see box on the **Production of selenomethionine labeled proteins**).

Production of selenomethionine labeled proteins (SeMet)

To exploit the anomalous scattering signal of an element (§4.3 and §11.2), a common approach is to introduce selenium atoms directly into the protein sequence. Indeed, the anomalous signal of selenium is located at the 0.98 Å wavelength, which is easily accessible on a synchrotron beamline. Selenium has chemical properties similar to those of sulphur, making it possible to replace the methionines in the proteins with selenomethionine (SeMet) without affecting its structural properties.

The first approach to replacing methionines with SeMet is to use a bacterial strain (*Escherichia coli*) that is auxotrophic for methionine, *i.e.* unable to synthesise methionine itself, as the host cell. This historic approach is simple to set up and allows almost 100% selenium substitution. For production, the host cell is placed in a flask containing a growth medium supplemented with SeMet. The cells mass produce the protein of interest by incorporating SeMet instead of methionines. However, the protein yield is often only 15%–20%, and it takes longer to grow *E. coli* in a minimal medium than in a standard medium. Alternatively, non-auxotrophic strains can be used in which methionine biosynthesis is inhibited prior to induction. With this approach, SeMet incorporation is around 90% with a higher protein yield.

Instead of *E. coli* bacteria, baculovirus-infected insect cells or yeast cells can be used, but the cytotoxicity of SeMet often results in only about 50% selenium incorporation. Mammalian cell lines can also be used as expression hosts and adapted for the production of SeMet-labelled proteins. Depending on the cell type the incorporation can vary from 60% to 90%.

References:

Hendrickson W.A., Horton J.R., LeMaster D.M. (1990) *EMBO J.* **9**, 1665–1672.

Doublié S. (1997) *Methods Enzymol.* **276**, 523–530.

Walden H. (2010) *Acta Crystallogr D Biol Crystallogr.* **66**, 352–357.

2.3 Purification

In most expression methods, the target protein is present in the host cells and must be extracted and purified. First, the sample is centrifuged to recover the cells, which are then resuspended in a buffer solution. The cells are then lysed to release their contents into the lysis buffer (figure 2.8). The cell walls and other

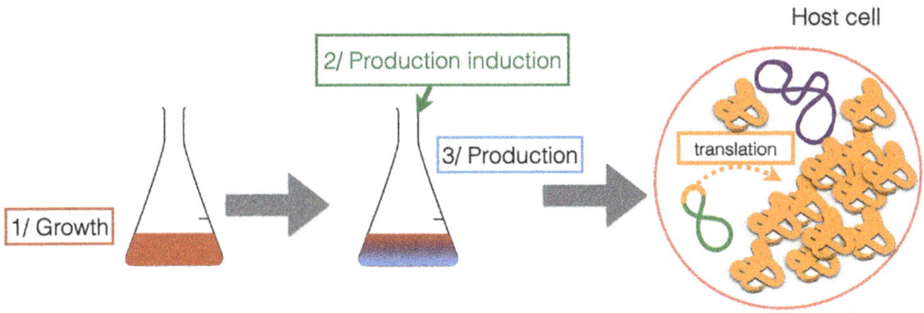

FIG. 2.7 – Production of the target protein by the host cell.

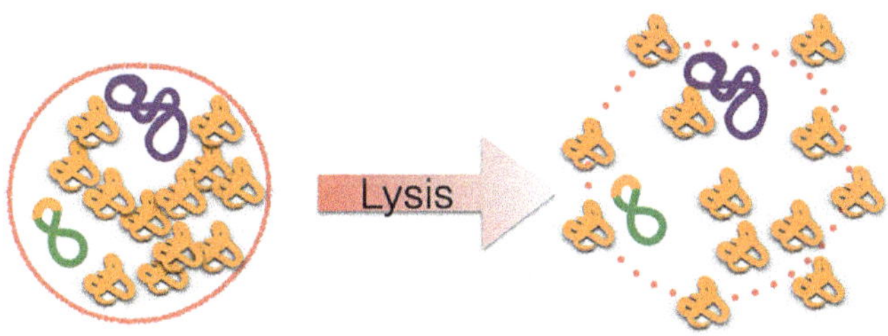

FIG. 2.8 – Lysis of cell walls and release of cell contents.

insoluble compounds form a pellet during the subsequent centrifugation and can be discarded. If the target protein is well soluble, it will remain in the soluble fraction, which also contains the other soluble proteins and compounds of the host cell.

The tag and cleavage site facilitate the purification process. The tag should resemble the components present in the cells as little as possible. It is chosen for its interaction properties during purification. In fact, a tag-specific affinity column is used first, *i.e.* a column to which the tag bound to the target protein will preferentially bind. The supernatant from the lysis step is injected into the column, and any compounds that are unable to interact with this column will pass into the filtrate (called 'flow through') (figure 2.9, 1). The target molecule sticks to the column thanks to its label (figure 2.9, 2). To recover the target protein, a compound is added that strongly interacts with the tag and displaces it from the affinity column by competition (figure 2.9, 3). The target is then recovered in the eluate in a single step (figure 2.9, 4).

At this stage, the purity of the protein solution should be around 80%. The target protein is still attached to the tag, which is usually removed before further structural biology studies. This is done by incubating the eluted protein with a highly specific protease capable of cleaving the protein construct at a specific site. After proteolysis, the solution contains the target protein cleaved from its tag, the tag, and the protease. To facilitate further purification, a common trick is to use a protease construct containing the same tag used for the target protein in the first step. By injecting the mixture containing the target protein, the separate tag and the tagged protease onto the same affinity column as before, the tag alone and the

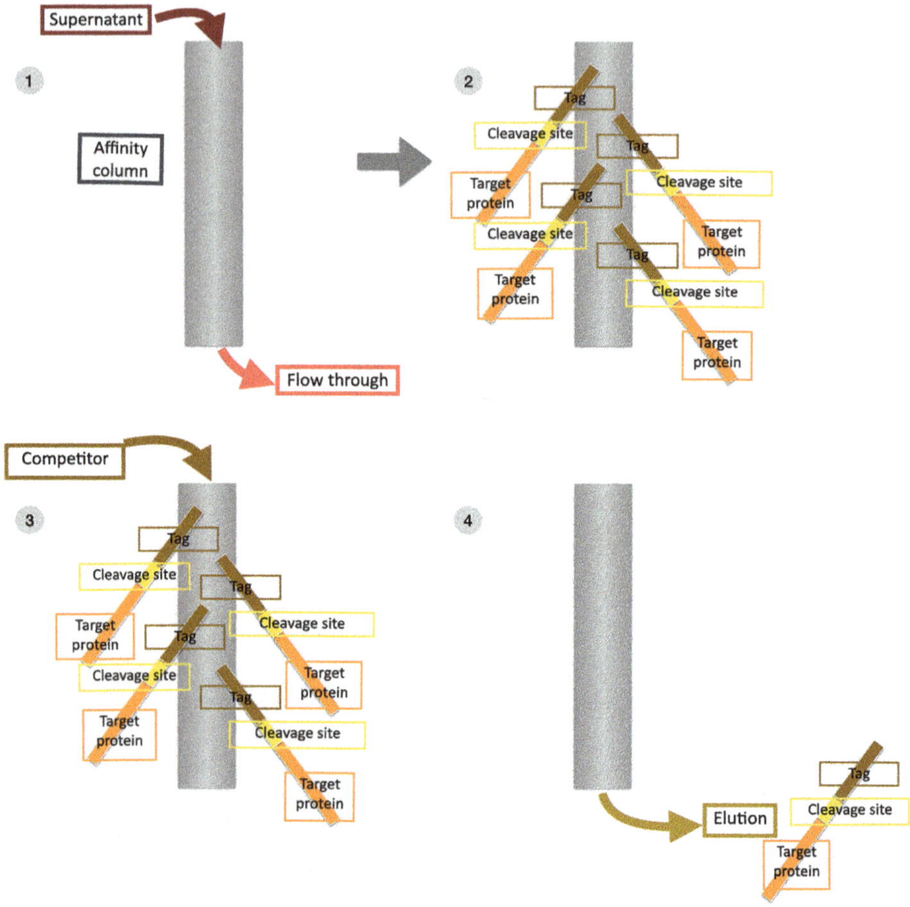

FIG. 2.9 – Affinity column purification steps.

tagged protease remain attached. In contrast, the target protein is released directly into the eluate as it no longer contains the tag (figure 2.10).

The contaminants retained during the first run on the affinity column are proteins that interact with the affinity column and are eventually eluted along with the target protein. During the second run on the affinity column, these contaminants are again retained. As it is no longer tagged, the protein is not retained on the column, and the impurities can be separated from the target protein. After this second run, the protein solution is typically 90%–95% pure. These two purification steps result in a highly homogeneous sample. In addition to these two purification steps, further polishing steps can be carried out by taking advantage of other properties of the target molecule such as its size, mass, ionic charge, etc.

It is then essential to check the quality and integrity of the sample and to concentrate it to levels suitable for the crystallisation experiments.

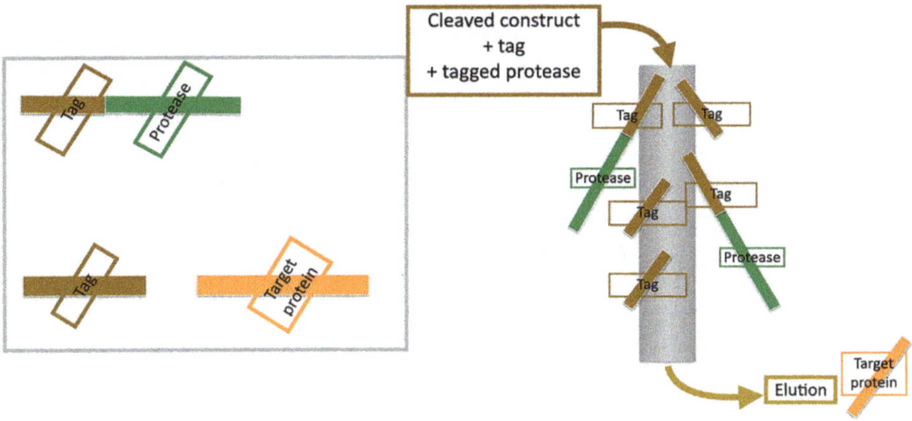

FIG. 2.10 – Left: Solution mixture after separation of the target protein and its tag with a specific protease; right: Separation of the target protein on a tag-specific affinity column.

Related videos

Prerequisites: knowing your sample (Les prérequis : connaître son échantillon)
https://youtu.be/rTYO0hsUGiU

Cloning, production, purification of samples (Clonage, production, purification des échantillons)
https://youtu.be/RhBaWJjxFlw

Visiting the laboratory: sample preparation (Visite du laboratoire : la préparation des échantillons)
https://youtu.be/ZA-nCSOloXE

Chapter 3

Crystal Features and Properties

Whether it is made up of small molecules such as chlorine and sodium atoms that form salt crystals, water that forms snowflakes, or macromolecules such as proteins, nucleic acids, or their assemblies, a crystal is a solid. The components of the crystal are always arranged in a regular manner, which gives them the properties essential for diffraction.

3.1 The Crystal Assembly

To understand the properties of the crystal assembly, let's start with a game of wallpaper. Figure 3.1 shows a wallpaper consisting of repeated patterns.

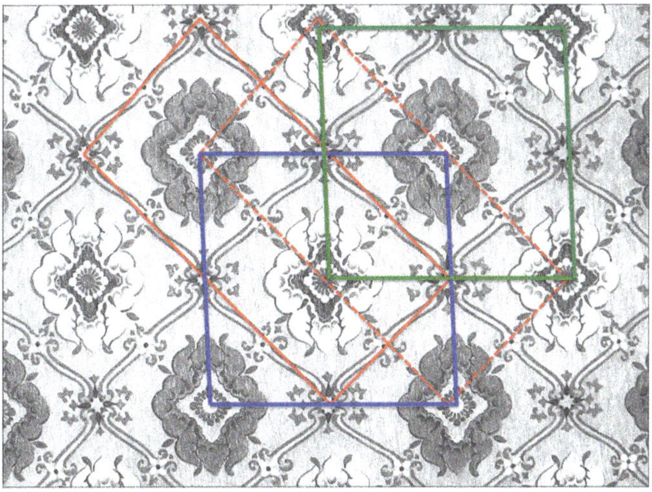

FIG. 3.1 – Examples of minimal units for describing whole repetitive designs by simple translation (adapted from https://commons.wikimedia.org/wiki/File:Un_motif_de_papier-peint_(Palais_An_Dinh,_Hu%C3%A9)_(4393490610).jpg).

DOI: 10.1051/978-2-7598-3501-0.c003

The minimum unit, the unit cell, can be represented as a square or a rectangle. The wallpaper is completely described by the repetition of this minimal unit in both directions of the figure. We can see that several positions of the same square describe the wallpaper (the blue square or the green square) and that the minimum unit can be contained in a quadrilateral of different shapes (the blue square or the green square compared to the red rectangle). The minimal unit always contains the same pattern, so it is up to us to choose one or the other to describe the wallpaper.

Similarly, a three-dimensional crystal can be described as the regular repetition along three directions of a box, the crystal unit cell, described by its cell parameters, and its symmetry operations or symmetries. A symmetry operation is a movement about a geometric element, which can be a point, a line or a plane. When we apply a symmetry operation to an object, we obtain an image of that object that is indistinguishable from the original, it is invariant. This transformation leaves the crystal unchanged. The cell parameters describe the length of the three sides a, b, and c and their angles alpha, beta, and gamma. A lattice often contains the same group of atoms more than once. Within the lattice, basic symmetry operations such as rotations and translations can often be used from one group of atoms to another. The asymmetric unit is the minimal group of atoms that allows the whole crystal to be reconstructed using crystal symmetry operations. Each asymmetric unit can contain one or more molecules, identical or not (figure 3.2).

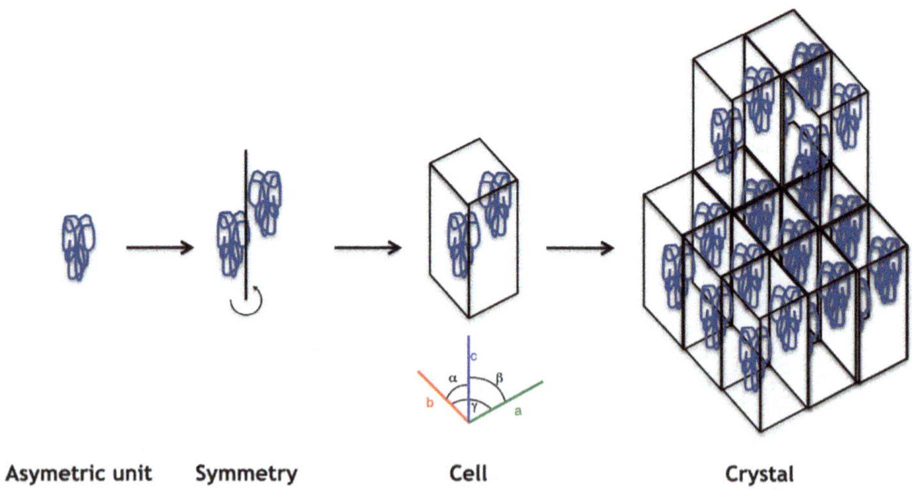

FIG. 3.2 – Crystal assembly from the asymmetric unit to crystal.

While the average size of a protein crystal is about 100 μm on a side, its typical cell length is about 100 Å or 10 nm. This makes 10 000 cells per side of a crystal or 10^{12} or 1000 billion cells in a crystal. The intensity of a Bragg spot is therefore proportional to the square of the number of cells (see §4.2). In addition, the interference produced by the crystal lattice is a powerful tool for concentrating the signal scattered by these molecules in certain directions.

3.2 Crystal Symmetries

The symmetry operations of a crystal are either rotations about an axis, reflections with respect to a plane, translations or combinations of all three. They are referenced in the International Crystallography Tables (see references in the Further Reading section). In biology, the symmetries observed are limited by the nature of biological macromolecules. They are chiral, which means that their image cannot be superimposed in a plane mirror (figure 3.3). Proteins are levogyres and the sugars in nucleic acids are dextrogyres. Therefore, for a given protein or nucleic acid, we cannot observe the mirror molecule: biological crystallography has no reflection symmetry.

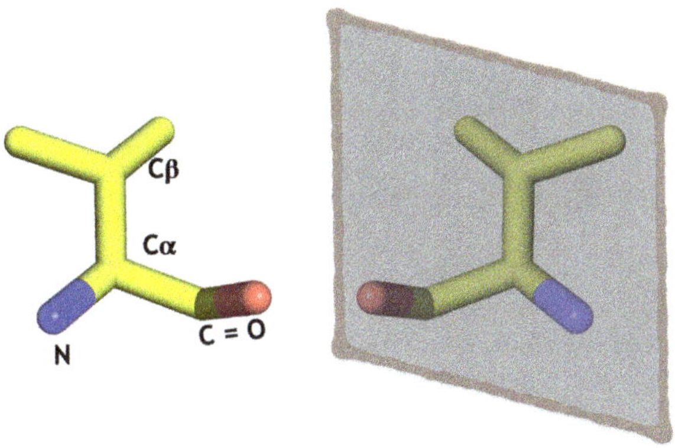

FIG. 3.3 – The chirality of an object refers to the fact that it cannot be superimposed on its image in a plane mirror.

Among the symmetries observed in biological crystallography are rotation symmetries about an axis. For a two-fold rotation, the molecule is rotated 180° around the axis. For a three-fold axis, the molecules are distributed every 120°, which corresponds to 3 molecules in 360°. There are also four-fold and six-fold rotations. These rotations can be combined with a translation along the axis of rotation. This is known as a helical translation or a screwed axis of rotation. For example, 2_1-symmetry occurs when a two-fold rotation about an axis is combined

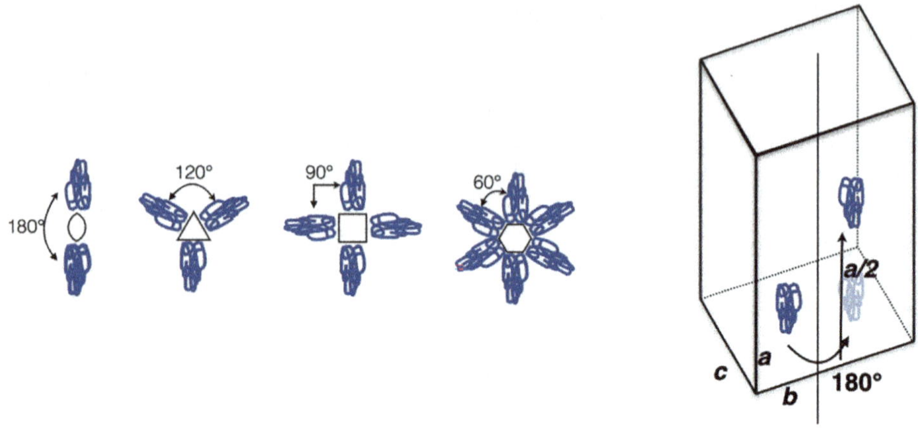

FIG. 3.4 – Examples of two-, three-, four- and six-fold axes; and two-fold screw axis 2_1.

with a translation of a distance equal to half the height of the cell (figure 3.4). This is similar to a half step around the axis of a screw. Similarly, a 3_1-symmetry corresponds to a rotation of 120° followed by a translation of $a/3$, and an axis 3_2-symmetry to a rotation of 120° followed by a translation of $2a/3$. It is also possible to describe four-fold or six-fold screw axes.

These symmetry operations can be found along any of the a, b or c axes. The symmetry operations of a given crystal define what is known as its space group. Space groups are also described in the International Crystallographic Tables.

3.3 The Bravais Lattices

We have seen that crystal results from the repetition of the unit cell in the three directions of the space. To describe this repetition, we use the concept of a lattice, which is a set of regularly arranged points, called nodes, whose positions are given by the expression: u**a** + v**b** + w**c**. Here u, v and w are integers and **a**, **b** and **c** are the vectors that form the basis of the lattice, whose length (a, b, c) and interaxial angles (α, β, γ) form the lattice parameters, as defined above.

In some cases, it is possible to identify certain relationships between the parameters or between the angles α, β, γ of the cell. For example, one can have $a = b$, and $\alpha = \beta = \gamma = 90°$. It is shown that there is a limited number of possible lattices: these are the 14 Bravais lattices (figure 3.5). These 14 Bravais lattices allow us to define 7 crystal systems corresponding to the different symmetries of the lattices. Each crystal system is characterised by specific constraints on the lattice parameters. For example, in the cubic system, all the parameters are equal and the angles are 90°.

Figure 3.6 shows the different types of unit cells, with the black dots representing the nodes of the lattice. Cells containing only one node are called primitive and are denoted P. To simplify the description of the lattice, it may be necessary

Cell parameters	Crystal system	Bravais lattice
$a \neq b \neq c$ $\alpha \neq \beta \neq \gamma \neq 90°$	Triclinic	aP
$a \neq b \neq c$ $\alpha \neq 90°;\ \beta = \gamma = 90°$	Monoclinic	mP, mC
$a \neq b \neq c$ $\alpha = \beta = \gamma = 90°$	Orthorhombic	oP, oF, oI, oC
$a = b \neq c$ $\alpha = \beta = \gamma = 90°$	Tetragonal	tP, tI
$a = b = c$ $\alpha = \beta = \gamma = 90°$	Cubic	cP, cF, cI
$a = b = c$ $\alpha = \beta = \gamma \neq 90°$	Rhomboedral	hR
$a = b \neq c$ $\alpha = \beta = 90°;\ \gamma = 120°$	Hexagonal	hP

FIG. 3.5 – The 14 Bravais lattices and their geometric characteristics.

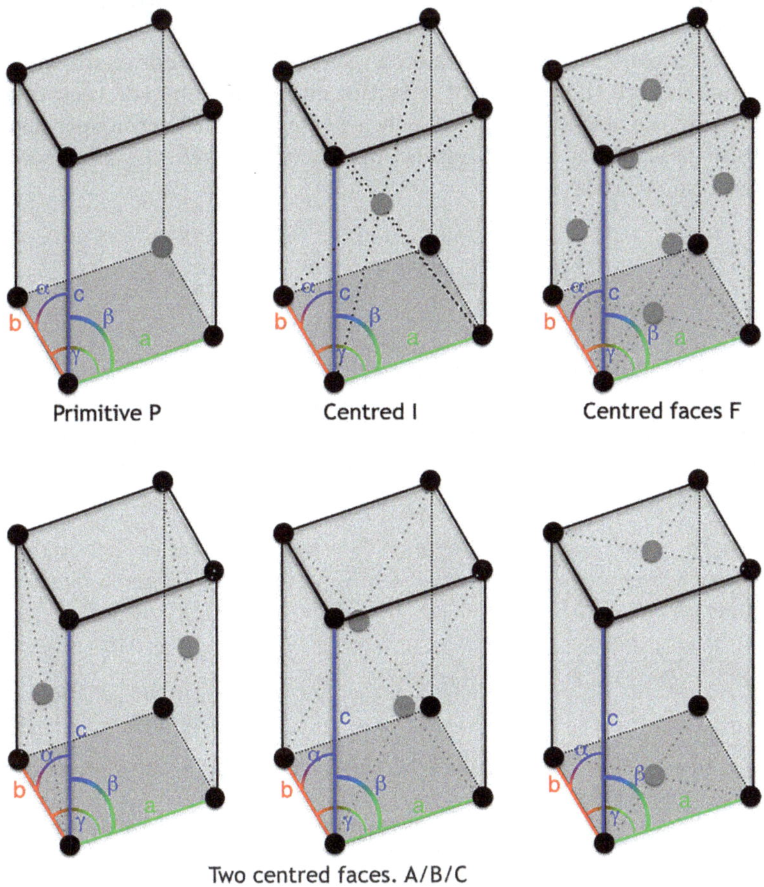

FIG. 3.6 – Representation of the different possible cell types.

to use multiple cells containing multiple nodes per cell. These lattices are conventionally called:

- centred, I, if the extra node is in the centre of the lattice;
- all-face centred, F, when the extra nodes are in the centre of the faces;
- face-centred, A, B or C, if the extra nodes are in the centre of the (b, c), (a, c) or (a, b) faces, respectively.

In the following (chapter 8) the content of the cell is characterised by its average electron density at the point x**a** + y**b** + z**c**, $\rho(xyz)$, where x, y and z are the fractional coordinates of the point (*i.e.* between 0 and 1).

3.4 The Reciprocal Lattice

It is possible to draw planes that pass through the nodes of the lattice, called lattice planes. These planes are essential in crystallography. We will see later on that the phenomenon of diffraction can be understood in terms of a reflection on these planes, which allows one to obtain the Bragg relation (§4.2). In crystallography, we call a 'family of lattice planes' the set of equally spaced parallel lattice planes containing all the nodes of the lattice.

To describe a plane, we specify its orthogonal direction. For example, the lattice plane passing through the origin of the lattice and defined by the vectors **a** and **b** is characterised by the vector product **a** ∧ **b** (see the box on **Vector operations**). This can be easily generalised to define another lattice, called the **reciprocal lattice**, whose three basis vectors are[1]:

$$\boldsymbol{a}^* = \frac{\boldsymbol{b} \wedge \boldsymbol{c}}{V};\ \boldsymbol{b}^* = \frac{\boldsymbol{c} \wedge \boldsymbol{a}}{V};\ \boldsymbol{c}^* = \frac{\boldsymbol{a} \wedge \boldsymbol{b}}{V}$$

where V is the volume of the cell given by $V = \boldsymbol{a} \cdot (\boldsymbol{b} \wedge \boldsymbol{c})$.

By convention, the positions of the nodes in this lattice are written as $h\boldsymbol{a}^* + k\boldsymbol{b}^* + l\boldsymbol{c}^*$, where the three integer indices h, k and l are the Miller indices. The vectors of the reciprocal lattice are expressed in Å^{-1}. This lattice has some very useful properties:

- The vector $h\boldsymbol{a}^* + k\boldsymbol{b}^* + l\boldsymbol{c}^*$ is orthogonal to a family of planes, denoted (hkl).[2]
- The distance between two consecutive planes, written d_{hkl}, is given by the inverse of the modulus of $h\boldsymbol{a}^* + k\boldsymbol{b}^* + l\boldsymbol{c}^*$.

[1]Alternatively, these vectors can be defined by multiplying them by 2π, which complicates the interreticular distance formulas.

[2]A small difficulty: this family of planes intersects all the lattice points and only them if the vector ha* + kb* + lc* is the smallest in its direction. For example, (080) represents planes 8 times closer than the (010) lattice planes...

Vector operations

Vectors are mathematical objects used to represent physical quantities that have a magnitude and a direction, such as speed, electric field or the basic translations of a lattice. They are usually written in bold: $\boldsymbol{v}$, $\boldsymbol{E}$, $\boldsymbol{a}$, $\boldsymbol{b}$ and $\boldsymbol{c}$, and their magnitude, or length, in light: v, E, a, b and c.

Simple operations can be performed on vectors. They can be added together, as shown here:

$\boldsymbol{a}+\boldsymbol{b}=\boldsymbol{d}$

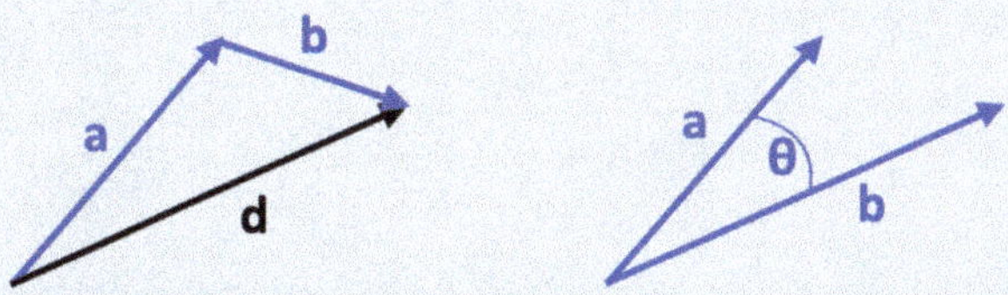

It is also interesting to introduce multiplication operations. The scalar product of two vectors $\boldsymbol{a}$ and $\boldsymbol{b}$ forming an angle θ is a number – a scalar – written $\boldsymbol{a}\cdot\boldsymbol{b} = \mathbf{ab}\cos(\theta)$. If the vector $\boldsymbol{b}$ is a unit vector, the scalar product $\boldsymbol{a}\cdot\boldsymbol{b}$ is the projection of $\boldsymbol{a}$ onto the direction of $\boldsymbol{b}$. If the two vectors are orthogonal, their scalar product is zero.

The cross or vector product of two vectors $\boldsymbol{a}$ and $\boldsymbol{b}$ is another vector denoted $\boldsymbol{a} \wedge \boldsymbol{b}$, orthogonal to a and b, of magnitude $|\boldsymbol{a} \wedge \boldsymbol{b}| = \mathbf{ab}\sin(\theta)$, such that $(\boldsymbol{a}, \boldsymbol{b}, \boldsymbol{a} \wedge \boldsymbol{b})$ is in the direct direction. The magnitude of this vector is equal to the area of the parallelogram defined by $\boldsymbol{a}$ and $\boldsymbol{b}$. The vector product is used in the definition of the basis vectors of the reciprocal space.

These two products allow us to define a third one: the scalar triple or mixed product. The mixed product of three vectors $\boldsymbol{a}$, $\boldsymbol{b}$ and $\boldsymbol{c}$ is a scalar: $(\boldsymbol{a}, \boldsymbol{b}, \boldsymbol{c}) = \boldsymbol{a}\cdot(\boldsymbol{b} \wedge \boldsymbol{c}) = (\boldsymbol{a} \wedge \boldsymbol{b})\cdot\boldsymbol{c}$. It is equal to the volume of the parallelepiped formed by these three vectors. It is therefore the volume of the cell of the base vectors $\boldsymbol{a}$, $\boldsymbol{b}$ and $\boldsymbol{c}$.

For example, in all types of crystal, the family of lattice planes defined by the vectors **a** and **c** is orthogonal to the vector $\boldsymbol{b^*}$ and written (010) and $d_{010} = 1/b^*$, the family of lattice planes defined by the vectors $\boldsymbol{b}$ and $\boldsymbol{c}$ is orthogonal to the vector $\boldsymbol{a^*}$ and written (100) and $d_{100} = 1/a^*$, ... (figure 3.7).

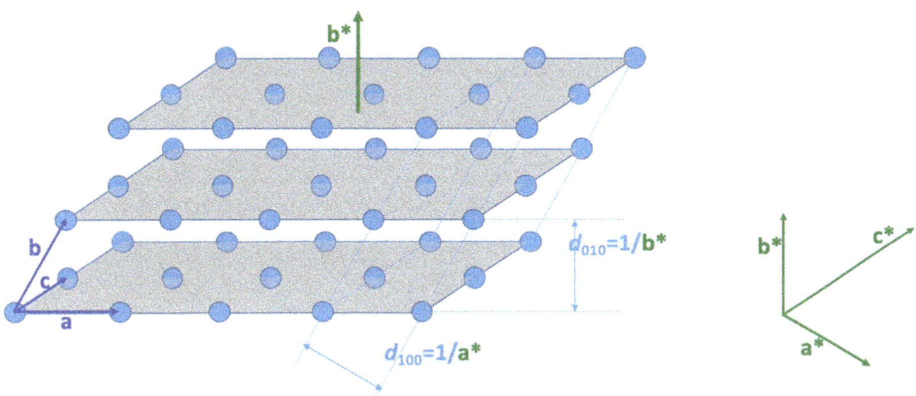

FIG. 3.7 – Example of lattice planes and associated reciprocal quantities for a hexagonal lattice. Note that this system, $\boldsymbol{b}^*$ is not parallel to $\boldsymbol{b}$ and $b^* = \frac{\sqrt{3}}{2b}$.

Related videos

The characteristics of a crystal and Bragg's law (Les caractéristiques d'un cristal et la loi de Bragg)
https://youtu.be/YjKhznck8f0

Chapter 4

X-Rays and Diffraction

4.1 Why X-Rays: The Interaction of Light and Matter

To see an object, you need light. Natural light, such as sunlight, or artificial light. But depending on the type of light, you may not see the same thing. The smallest details you can see depend directly on the colour of the light used. The colour of light is characterised by its frequency ν, or its wavelength in vacuum λ, which are related by the formula $\lambda = c/\nu$ (c: the speed of light in vacuum). The colours of the rainbow range from violet to dark red in the visible spectrum. The corresponding wavelengths are 400 nm for violet and 700 nm for dark red. Beyond dark red is infrared radiation, with wavelengths between 1 μm and 1 cm, and then radio waves, with wavelengths between 1 m and 1 km. The shorter wavelengths are ultraviolet light up to 100 nm, X-rays between 10 nm and 0.1 nm and gamma rays with very short wavelengths of 0.01 nm or less.

The wavelength of the light used to illuminate an object determines the minimum distance that can be observed: this minimum distance is approximately half

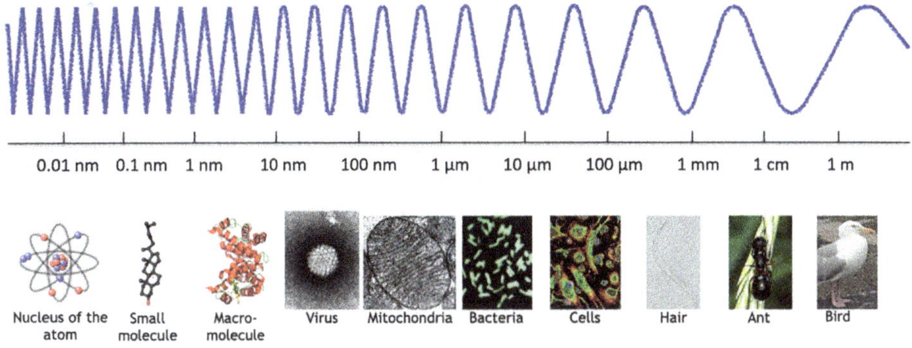

FIG. 4.1 – The size of the objects that we can observe depends on the wavelength of the light used to illuminate them (ant: https://commons.wikimedia.org/wiki/File:Formica_high_res.jpg; cells: https://commons.wikimedia.org/wiki/File:Epithelioid_cells_and_macrophages_in_cell_culture.jpg; bacteria: https://commons.wikimedia.org/wiki/File:Dark_field_microscopy_revealing_Shigella_dysenteriae_bacteria.jpg; virus: https://commons.wikimedia.org/w/index.php?curid=1097459).

DOI: 10.1051/978-2-7598-3501-0.c004

the wavelength (see box on the **Resolution limit and wavelength of light**). Therefore, the shorter the wavelength, the smaller the objects that can be observed (figure 4.1).

4.1.1 The Choice of X-Rays

Whether it is a small molecule or a biological macromolecule, the distances associated with chemical bonds are in the order of 0.1–0.18 nm. This corresponds to 1–1.8 Å in the usual crystallographic units (figure 4.2).

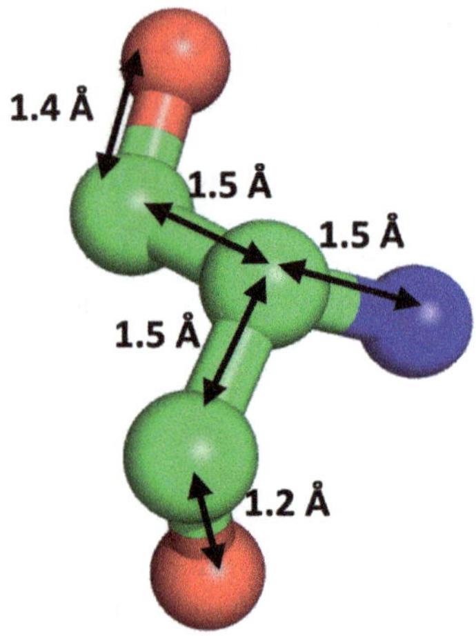

FIG. 4.2 – Representation of an organic molecule and the distances between the atoms. The atoms are shown in green for carbon, blue for nitrogen and red for oxygen. This is the model used in crystallography: the hydrogens are not shown.

Consequently, to observe a molecular structure at the atomic scale, the wavelength of the light required is about 0.1 nm or 1 Å. This light corresponds to X-rays. However, our eyes cannot see X-rays and therefore cannot directly see an object illuminated by X-rays. Images are recorded using special cameras or detectors, coated with a compound or emulsion that is sensitive to X-rays. In a sense, these detectors see the X-ray photons and pass the information on to us.

4.1.2 The Interaction of Molecules with X-Rays

X-rays can interact with atoms in a number of ways (see box on the **Interaction process between light and atoms**). They can interact with the electrons of the atoms without any loss of energy and are then scattered in all directions – this is known as elastic scattering. The intensity of the scattered X-ray beam is a function of the number of electrons in the atom and the scattering angle. To account for this scattering, a scattering factor, f, is introduced for each atom in the periodic table (see the box on the **Atomic scattering factor**).

Interaction process between light and atoms

In a crystallography experiment, X-ray photons penetrate the crystal and can interact in different ways with the atoms they encounter. To illustrate these different types of interactions, consider a protein crystal a few tens of μm in size, illuminated by X-rays of 1 Å wavelength (*i.e.*, 12.4 keV photon energy). In this situation, the vast majority of photons, about 98%, pass through the crystal without any interaction, as if the crystal were transparent. For the 2% of photons that do interact:

- 1.74% are absorbed by atoms in the crystal and transfer their energy to an electron in the atom. This is the photoelectric effect, which results in the production of high-energy electrons, the main cause of crystal degradation in the experiment.
- 0.14% are scattered inelastically, transferring a small amount of their energy to the atoms. This is the Compton effect, which involves a change in the wavelength of the scattered X-rays. The scattered photons contribute to the diffuse background in the diffraction pattern.
- 0.12% are scattered elastically, with no change in wavelength, and only these can give rise to the interference that is the basis of diffraction.

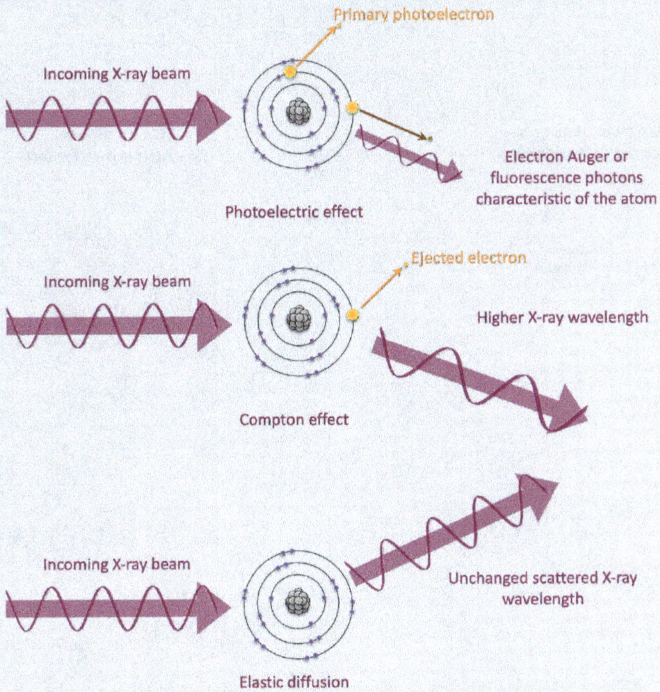

This distribution depends on both the energy of the X-rays and the chemical composition of the crystal. The presence of heavier, electron-rich atoms increases the proportion of absorption (photoelectric effect). The use of higher energy X-rays reduces the probability of interaction unless the crystal contains heavy, *i.e.* electron-rich, atoms and this energy is close to one of the absorption edges of these atoms.

Now consider the case of two atoms. Each atom receives and scatters the same X-rays. However, the amplitude of the wave varies according to the direction of scattering. In fact, the scattered wave is the superposition of the waves scattered by the two sources, *i.e.* the two atoms. This is the phenomenon of interference. In some directions the interference produces a maximum amplitude, this is called constructive interference (figure 4.3a). In other directions the interference produces a zero signal, this is called destructive interference (figure 4.3b). To simplify how constructive or destructive interference occurs, consider what happens to each atom. Each atom receives the same beam and scatters it. In some directions, the maxima of the scattered waves match. The scattered waves are in phase. The addition of the waves is maximal, the interference is constructive (figure 4.3c). In other directions, the maximum of the scattered wave amplitude matches with the minimum of the other, the scattered waves are in antiphase, the interference is destructive and the resulting wave amplitude is zero (figure 4.3d). Between these two extremes, the intensity varies between a zero signal and a signal equal to the sum of the two amplitudes.

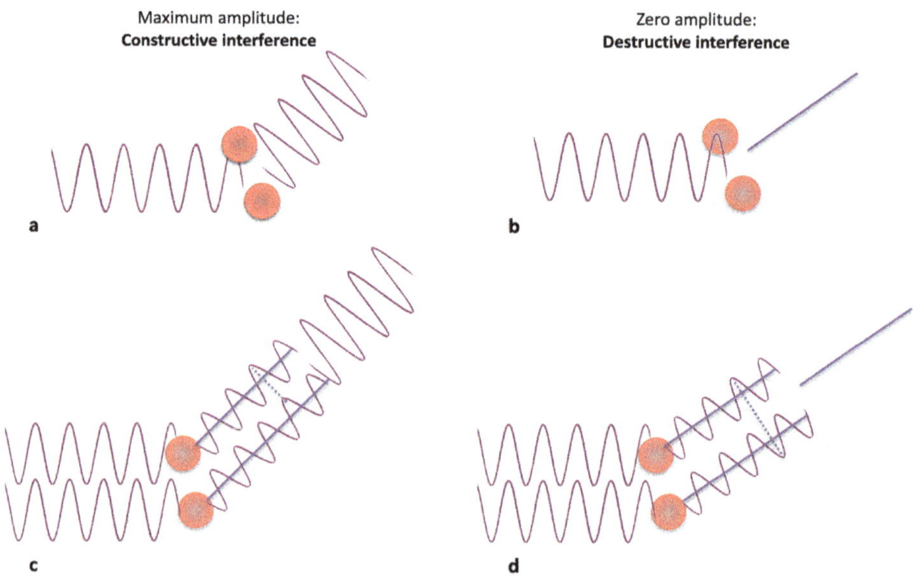

FIG. 4.3 – Illustration of the phenomenon of constructive (a and c) or destructive (b and d) interference in the case of two atoms.

Atomic scattering factor

Consider the scattering of a wave by an isolated atom. The incident wave has a wave vector $\boldsymbol{k_i}$ ($k = 2\ \pi/\lambda$) and the scattered wave has a wave vector $\boldsymbol{k_d}$ (see figure on the left). The process is characterised by the scattering vector $\boldsymbol{q} = \boldsymbol{k}_d - \boldsymbol{k}_i$. The atomic scattering factor $f(\boldsymbol{q})$ gives the elastic scattering amplitude for an incident wave of unit amplitude. It is shown to be equal to the Fourier transform of the electron density $\rho(\boldsymbol{r})$ of the atom:

$$f(\boldsymbol{q}) = \int \rho(\boldsymbol{r})\, e^{2i\pi\boldsymbol{q}\cdot\boldsymbol{r}}\, d\boldsymbol{r}$$

A classical interpretation allows us to understand the physical meaning of this result, even though it can only be obtained rigorously from quantum mechanics. The charges $\rho(\boldsymbol{r})d\boldsymbol{r}$ in each volume element $d\boldsymbol{r}$ of the electron cloud begin to vibrate under the action of the electric field of the incident wave (see figure, left).

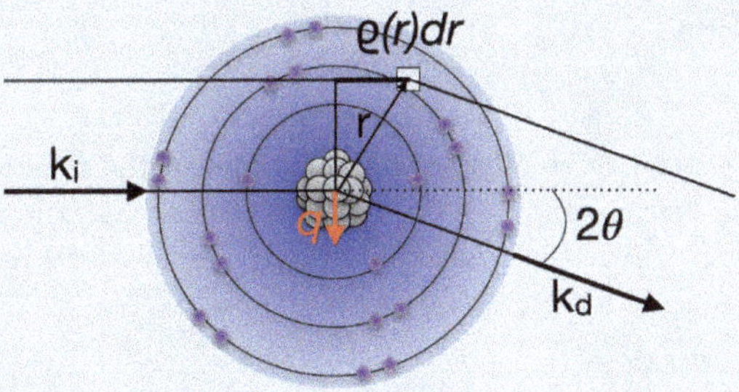

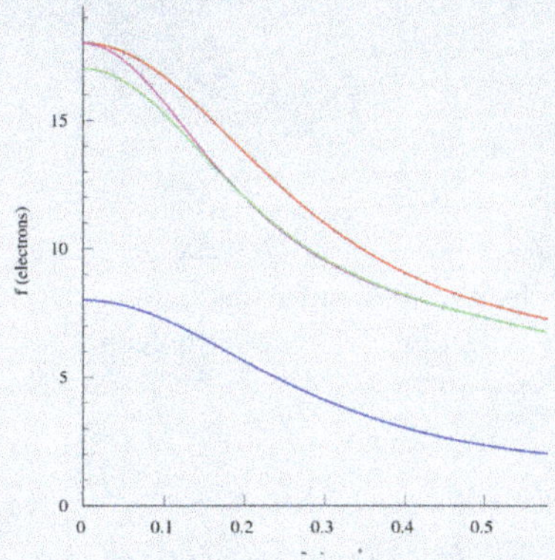

Left, diagram of single atom diffusion. Right, atomic scattering factor for oxygen (blue), chlorine (green), Cl^- anion (magenta) and K^+ cation (red) (https://fr.wikipedia.org/wiki/Facteur_de_diffusion_atomique#/media/Fichier:Cromer-Mann_structure_factors.svg).

Like little antennas, these elementary charges will re-emit wavelets in all directions, which will add up and interfere to give a wave whose amplitude depends on the scattering factor $f(\boldsymbol{q})$ and therefore on the scattering angle 2θ (the scattering factor $\boldsymbol{q}$ depends on θ by the formula $q = 4\pi\frac{\sin\theta}{\lambda}$). The scattering factor is therefore an atomic shape factor which has the appearance of a bell curve (see figure on the right) for all atoms. Its value at the origin is equal to the number of electrons in the atom. This helps to understand why light elements scatter little and why two neighbouring elements have very close atomic factors, making them difficult to distinguish (see green, red and magenta curves).

Macromolecules are made up of hundreds or even thousands of atoms. The interference principle applies in the same way as for two atoms. Consequently, the signal resulting from the interaction of an X-ray beam with the molecule varies according to the number of electrons in the atoms that make up the molecule, but also according to the number and arrangement of the atoms in relation to each other. In other words, it depends on the three-dimensional structure of the macromolecule (figure 4.4).

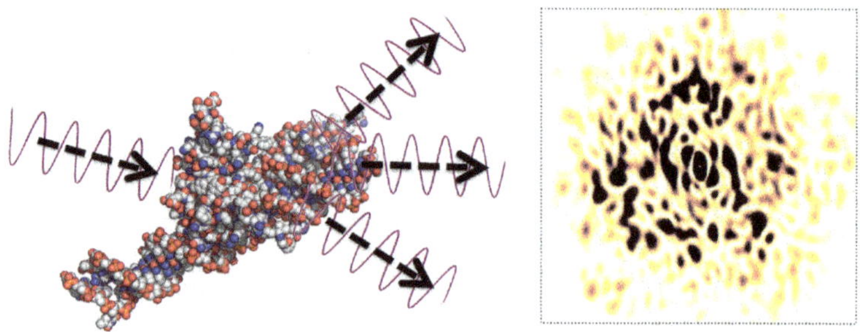

FIG. 4.4 – The scattering signal depends on the three-dimensional structure of the macromolecule. The scattering image was calculated with James Holton's nanoBragg software, from the pdb file of the macromolecule shown on the left of the figure (pdb: https://www.rcsb.org/structure/3UKG).

4.1.3 How X-Rays Interact with a Crystal Lattice

Now consider a large number of identical atoms arranged regularly in an ordered lattice. When an X-ray beam hits this lattice, hundreds of thousands of waves interfere with each other. Because the atoms are regularly spaced, the same phenomenon occurs with two atoms, repeated as many times as there are atoms in the lattice. In some directions the scattered waves are superimposed, they are in phase. Constructive interference occurs, the amplitudes add up and the resulting intensity, equal to the square of the amplitude, is very strong. On the contrary, as soon as we deviate from these specific directions, the interference is no longer constructive. The scattered intensity becomes negligible. As a result, the image observed on the detector reconstructs a regular pattern of points, called Bragg spots, corresponding to the constructive interferences. They form the so-called reciprocal lattice (figure 4.5, see §3.4).

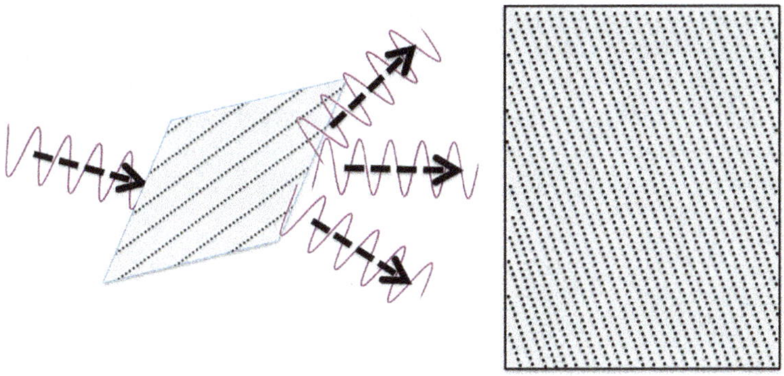

Fig. 4.5 – The constructive interference of a lattice forms a lattice called a reciprocal lattice.

4.2 Diffraction: Bragg's Law

A macromolecular crystal combines the properties of the macromolecule and the crystal lattice. When an X-ray beam illuminates a macromolecular crystal, the interference is constructive only in characteristic directions that depend on the crystal lattice. This produces the diffraction pattern, which is a distorted image of the reciprocal lattice and can be recorded on an X-ray-sensitive detector. Each point in the recorded diffraction pattern is called a Bragg spot. Its intensity is proportional to the square of the number of molecules in the crystal and depends on the organisation of the atoms in the molecule relative to each other, *i.e.* its three-dimensional structure. The intensity of each Bragg spot therefore contains part of the information about the overall structure of the macromolecule (figure 4.6).

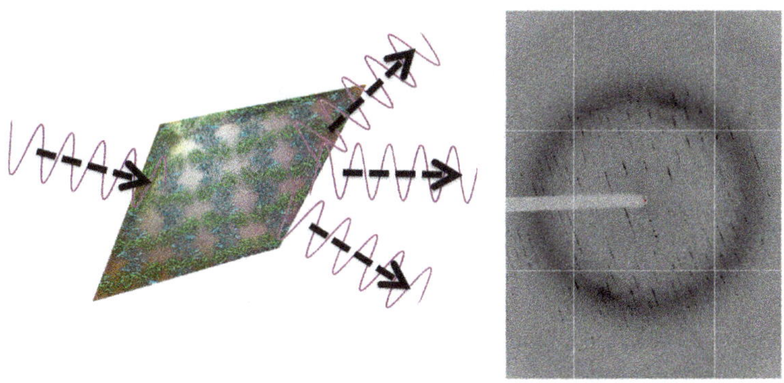

Fig. 4.6 – The molecules organised in a crystal form a regular lattice. Interference is constructive in discrete directions that depend on the crystal lattice.

Because of the regular organisation of the molecules in the crystal lattice, the atoms are packed into the crystal form layers called crystal or lattice planes. Diffraction occurs when X-rays are reflected from these crystal planes, as from a mirror. Two successive planes are separated by a distance *d*. Consider a beam of wavelength λ incident on the planes at an angle θ and scattered at the same angle θ. Depending on the distance *d* between the two planes and on the angle of incidence, there is an optical path difference between the waves reflected from the same family of crystalline planes. This difference is equal to twice the sine of the angle multiplied by the distance *d*, since the path of the outgoing wave must be taken into account (figure 4.7).

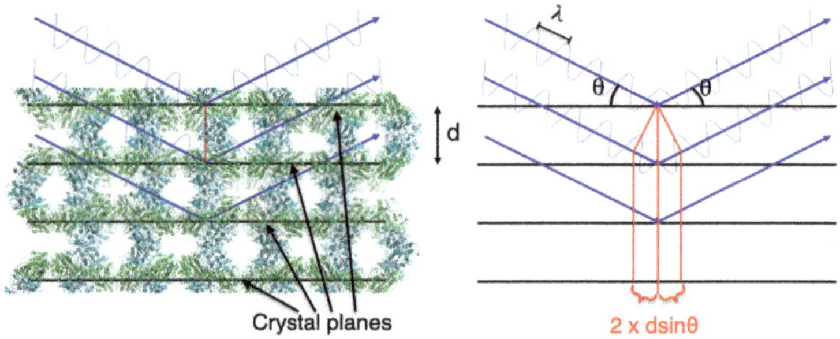

FIG. 4.7 – Constructive interference condition when a beam of light of wavelength λ is diffracted by crystal planes separated by a distance *d* at an angle of incidence θ.

The path difference between waves reflected from planes separated by a distance *d*, with an angle of incidence and reflection θ, is equal to 2-time $\sin\theta$ multiplied by the distance *d*, *i.e.* $2d\sin\theta$. Constructive interference occurs when this difference $2d\sin\theta$ is equal to the wavelength or an integer multiple n of the wavelength. We can write this as

$$n\lambda = 2d\sin\theta \tag{4.1}$$

This fundamental law is the Bragg's law. If there is one formula to remember, it is this one, which is constantly used at all stages of the resolution of a crystallographic structure. Each parameter is associated with an important element of the experiment:

- λ: the X-ray;
- *d*: the crystal;
- θ: the diffraction.

The angle 2θ corresponds to the scattering angle, *i.e.* the angle between the incident and the scattered beam. It increases as one moves away from the centre of a diffraction pattern. The angle $2\theta_{max}$ is the maximum angle at which reflections are observed (figure 4.8). For a given wavelength, this angle $2\theta_{max}$ determines the resolution limit of the data (see box on the **Resolution limit and wavelength**).

Resolution limit and wavelength

Since the wavelength λ is fixed, the application of Bragg's law is sufficient to determine the theoretical diffraction limit of a crystal. Knowing the distance from the crystal to the detector and the distance from the centre of the beam to the most distant points in the diffraction pattern, the angle $\theta_{\max}$ can be calculated. The distance $d_{\min}$ can be calculated as follows:

$$\lambda = 2d_{\min} \sin\theta_{\max}$$

Therefore: $d_{\min} = \lambda/2\sin\theta_{\max}$

The distance $d_{\min}$ is the smallest distance we can measure from the reflections at the largest angle $2\theta_{\max}$, and represents the size of the smallest details we can observe with these diffraction data. The angle $2\theta_{\max}$ therefore determines the maximum resolution of the diffraction pattern, in Å. We will see later how the maximum resolution limits the interpretation of the three-dimensional structure of the crystallised macromolecule. The diffraction geometry of a crystal can be easily visualised using a geometric construction called the Ewald sphere, which is a three-dimensional interpretation of Bragg's law.

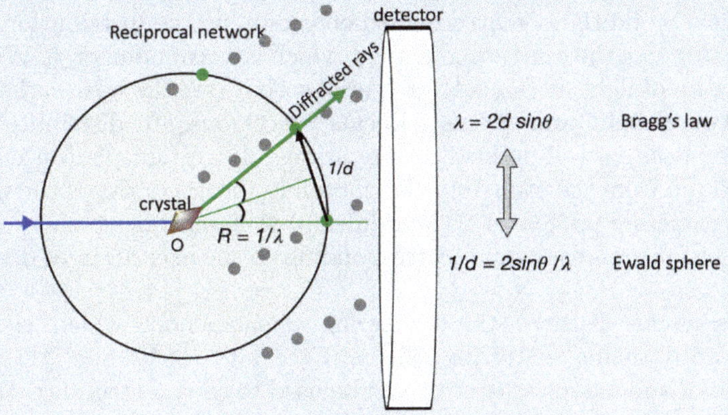

The radius R of the sphere is the inverse of the X-ray wavelength. The crystal is at the centre of the sphere and the reciprocal lattice is centred at the intersection of the sphere and the incident beam. Each node of the reciprocal lattice has a corresponding vector perpendicular to a family of reticular planes. For the node closest to the origin, the Miller indices h, k and l are mutually prime and the modulus of the corresponding vector is equal to the inverse of the interplanar distance. This construction shows that diffraction occurs only when a node of the reciprocal lattice intersects the Ewald sphere. In this case, a ray from the centre of the sphere (the crystal) is diffracted in the direction of the node of the reciprocal lattice that intersects the Ewald sphere.

In the above, we have assumed that the crystal is perfectly ordered and that all the cells are strictly identical. In reality, crystals are slightly disordered and the contents of the cells vary slightly, either because of thermal agitation, in which case the disorder is dynamic, or because the conformations of the molecules vary from one cell to another. In the directions where diffraction occurs, this disorder causes the rays scattered by the molecules to be out of phase and the intensity of the Bragg spots to decrease. This decrease in scattered amplitude is modelled by an exponential factor $\exp(-B(\frac{\sin\theta}{\lambda})^2)$ called the Debye–Waller factor, where B is a quantity characterising the degree of disorder. Thus, the greater the disorder, the greater the decrease in spot intensities at large angles. The maximum diffraction angle max decreases and so does the resolution.

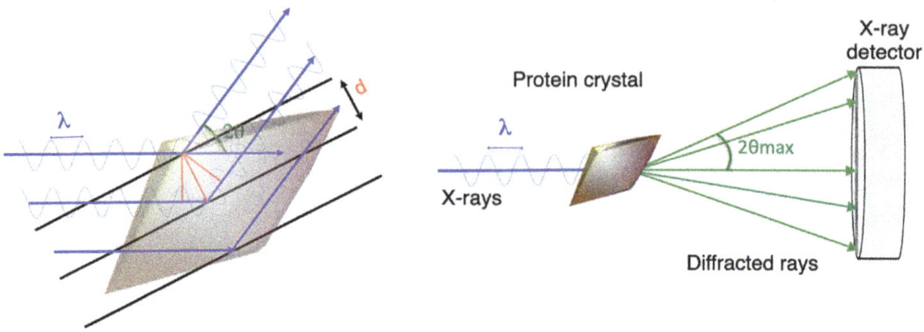

FIG. 4.8 – Bragg's law applied to the scale of a crystal.

4.3 Anomalous Diffusion

Under 'normal' conditions, scattering depends only on the distribution of electrons around the atom *via* the scattering factor f, which is a real number f_0. When an atom is hit by a beam of light, it can scatter it or absorb it (see the box on the **Interaction process between light and atoms**). Because electrons are distributed in atomic orbitals, they occupy well-defined energy states. At certain photon energies, electrons are ejected from the atom and the photon is absorbed: this is the photoelectric effect. The scattering process is also modified at these energies (or wavelengths): we observe a resonance phenomenon corresponding to the excitation of deep electronic layers.

This resonance disturbs the scattering phenomenon, which is then called anomalous or resonant scattering. We will see later (§7.1 and §11.2) that this phenomenon of anomalous scattering can be used to solve a structure, thanks to the possibility of tuning the wavelength on a synchrotron beamline.

Related videos

The characteristics of a crystal and Bragg's law (Les caractéristiques d'un cristal et la loi de Bragg)
https://youtu.be/YjKhznck8f0

Light/matter interaction (Interaction lumière/matière)
https://youtu.be/HiBrbKFsX4o

Chapter 5

Crystallisation of a Biological Macromolecule

5.1 General Principles

Starting from a protein in solution, crystallisation requires transient energy and time. With few exceptions, the solution must be very pure and concentrated. The initial concentration must be sufficient for the proteins to interact. Spontaneously, proteins interact randomly, leading to the formation of disordered or amorphous aggregates and eventually to protein precipitation. However, under favourable conditions,

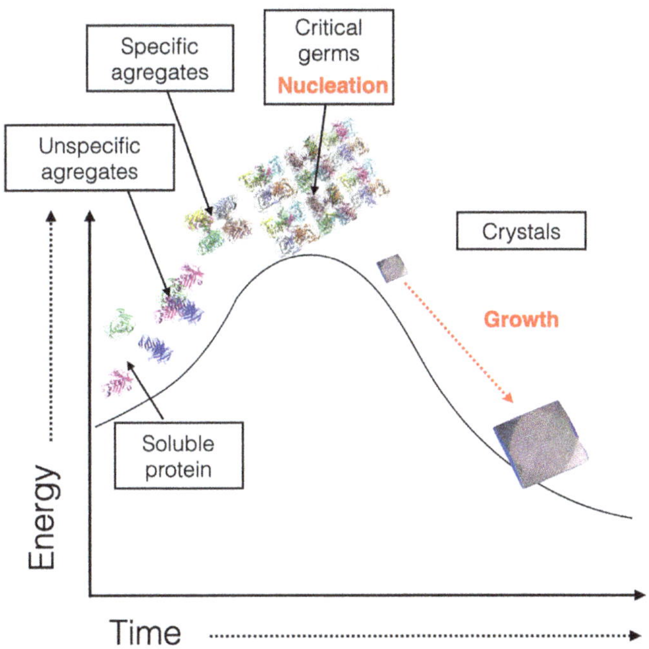

FIG. 5.1 – Energy barrier associated with the various steps leading to the formation of crystals of macromolecules.

DOI: 10.1051/978-2-7598-3501-0.c005

more specific interactions occur, leading to ordered aggregates that can develop into a seed crystal or critical seed: this is the nucleation stage. Crystal growth starts when the energy barrier associated with the formation of the critical seed can be overcome, and it stops when the maximum crystal size is reached (figure 5.1).

5.1.1 Properties of Proteins in Solution

Each protein has a unique composition and, therefore, specific properties. To obtain a homogeneous, ordered solid crystal, the molecules must be identical in chemical composition, physicochemical properties and structure. It is therefore essential not only to have a solution of high chemical purity but also to reduce as much as possible the flexible regions that affect the conformational homogeneity of the macromolecule and prevent the formation of stable intermolecular contacts. The difficulty is to find, for a given macromolecule, the favourable conditions that allow the formation of a critical ordered seed and subsequent crystal growth. To do this, it is necessary to test the many parameters that affect the solubility of the protein.

Proteins are polyelectrolytes. Their surface is covered with polar charges and groups that interact with the solvent molecules. Therefore, their solubility is influenced by the pH, the ionic strength, the temperature or the dielectric constant of the solution (figure 5.2, left). In particular, the protein's isoelectric point (pI), *i.e.* the pH at which the total net charge of the molecule is zero, is important because it corresponds to the pH at which protein solubility is minimal. To improve conformational homogeneity in solution, the addition of ligands or partners to the solution can stabilise a particular conformation of the protein and thus promote nucleation. If the protein contains disordered and mobile loops, it may also be useful to design mutants in which these regions are deleted or truncated (figure 5.2, centre). Finally, it may be necessary to crystallise an isolated domain of a large protein. However, the isolation of a domain may reveal hydrophobic regions on the surface of the protein,

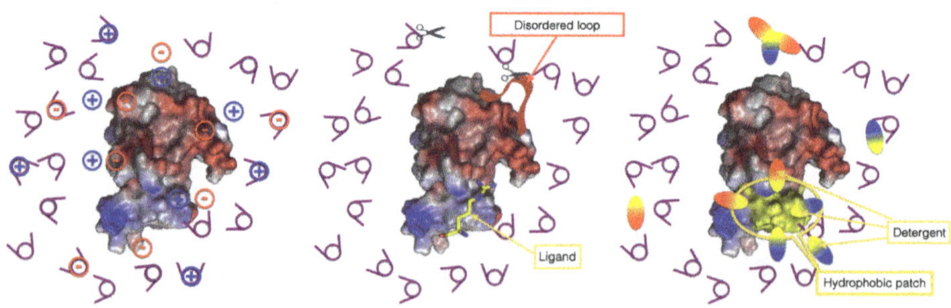

FIG. 5.2 – Properties of proteins that affect their solubility: polyelectrolyte on the left, ligand interaction zone and disordered loop in the middle, hydrophobic region and detergent effect on the right. Solvent molecules are shown in purple with their charge as + and − signs, and detergent molecules are shown as two-coloured ovals, the yellow part corresponding to the hydrophobic region, the blue or red part to the positively or negatively charged hydrophilic region.

which may affect solubility. The presence of detergents in the solution hides the hydrophobic regions and improves the solubility of these samples (figure 5.2, right). In the case of membrane proteins (see box on **Crystallisation of membrane proteins**), the use of detergents will be necessary because the protein interacts with the membrane *via* hydrophobic regions. Once these properties are known, a general scheme for the behaviour of the protein in solution can be proposed.

5.1.2 General Scheme of Protein Behaviour in Solution

In order for a protein to crystallise, it is necessary to bring it as close as possible to the limit of its solubility in the solution. The first essential parameter is the protein concentration, which must be high enough to favour interactions. Some chemicals, described below, affect the properties of proteins in solution. We will call them crystallisation agents. We can therefore plot a protein solubility curve as a function of the crystallisation agent concentration (figure 5.3). The area above the solubility curve is the supersaturation area. Just above the solubility curve there is a metastable zone where the solution is already supersaturated with the protein, but it remains clear: the proteins do not precipitate. In the precipitation zone, non-specific interactions dominate and proteins precipitate. The nucleation zone lies in an intermediate region between the metastable zone and the precipitation zone. As the crystal grows, it absorbs protein from the solution and the protein concentration gradually decreases. When the protein concentration in the solution reaches the solubility curve, crystal growth stops. An equilibrium is established between the incorporation of molecules into the crystal and the dissolution of the crystal (figure 5.3). This pattern is very general. For each new macromolecule, the nature and concentration of the agent that will enable crystallisation must be determined.

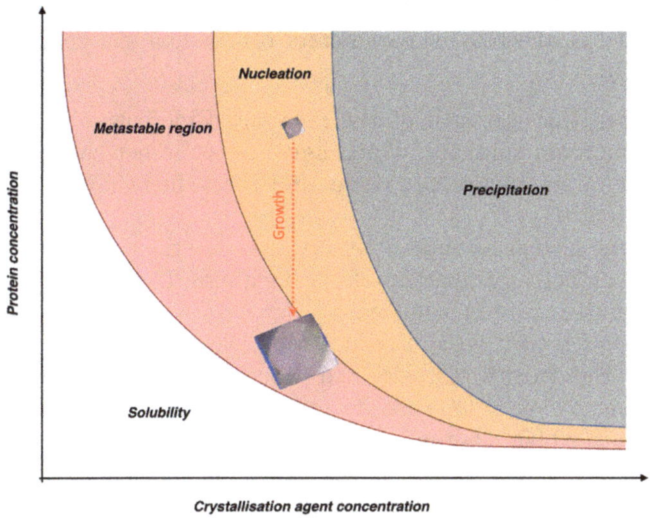

FIG. 5.3 – General scheme of nucleation and growth conditions.

Crystallisation of membrane proteins

Membrane proteins are proteins associated with the cell membrane. They are divided into two broad classes according to the nature of their interaction: 'peripheral' and 'integral'.
Peripheral membrane proteins do not cross the lipid bilayer of the membrane. They are tethered by electrostatic and hydrophobic interactions and can be dissociated by changes in pH or ionic strength. These proteins can be crystallised as soluble proteins.
Integral membrane proteins contain hydrophobic structural motifs (alpha helices or beta sheets) that enable them to cross and anchor to the cell membrane. Because of the hydrophobic transmembrane moiety, detergents are used to dissociate and solubilise these proteins. Crystallisation of a membrane protein is a challenging task and when crystals are obtained, they are often very small and difficult to visualise. There are two main crystallisation methods:(i) *In surfo* crystallisation uses micellar solutions of protein and detergent. The detergent micelles coat the hydrophobic regions of the protein. The protein is then crystallised as a soluble protein. The crystals formed in this way involve polar protein-protein interactions. They are very fragile due to the large solvent channels present in this type of stacking.
(ii) Liquid cubic phase (LCP) crystallisation is based on the incorporation of the protein into a lipid matrix consisting of 60% lipid and 40% protein, followed by the addition of a crystallising agent. The protein associated with the detergent diffuses into the lipid bilayer and crystal nucleation and growth occur as the system reaches equilibrium. The liquid cubic phase is not easily obtained and is only stable within certain temperature ranges. In addition, due to the extreme viscosity of LCP, droplet placement requires the use of special equipment and handling requires some practice. LCP droplets can be placed by vapour phase or batch diffusion, but the most common method uses a cyclic olefin copolymer (COC) sandwich. The COC offers many advantages as it prevents crystal dehydration, is optically transparent and absorbs and scatters X-rays very weakly.

5.1.3 Crystallisation Agents

Protein solubility is affected by parameters other than the concentration of the crystallising agent:

- Salts affect the ionic strength of the medium. Different salts, such as lithium sulphate or ammonium sulphate, which have the same net charge, may also have different protein-ion interaction properties and affect the behaviour of the protein in solution in different ways.
- Organic solvents, such as alcohols or ethers, decrease the chemical activity of water and reduce the dielectric properties of the medium. They also affect the density, viscosity and diffusion rate in solution.
- Polymers cause dehydration and reduce the dielectric properties of the medium. They also affect the density, viscosity and diffusion rate in the crystallisation solution. Polyethylene glycols (PEG) are the most commonly used polymers. They vary in size from 200 to 20 000 Daltons.

This list of parameters is not exhaustive. For each sample, it will be necessary to vary them one by one to find the crystallisation conditions.

5.2 Approaches, Platforms

The crystallisation of a macromolecule is difficult to obtain and reproduce due to the complexity of the mechanisms involved. For a given molecule, the use of screens allows us to investigate the conditions of crystal nucleation. The optimisation phase then aims to reduce the number of crystals per experiment in order to increase their size. To achieve this, crystallographers have developed a wide range of methods and empirical approaches. These methods are constantly evolving, and we present here an overview of the most classical ones.

5.2.1 Equipment Needed

The basic equipment consists of:

- A protein sample that is more than 95% pure and highly concentrated, approximately 10 mg/ml. As ionic strength is a variable used in the crystallisation experiments on a macromolecule, it is important to dialyse the sample against a low ionic strength buffer before starting a crystallisation experiment. It should also be filtered or centrifuged to remove any aggregates. Finally, depending on the properties of the molecule, specific stabilising agents are added to the solution.

- A temperature-controlled room where the experiments are carried out and the crystallisation plates are stored. Temperature is one of the parameters that modify the solubility of the protein.

- A microscope to observe the formation of the crystals.

- Chemical compounds of high purity to maximise the reproducibility of the experiments.

In order to study the parameters mentioned in the previous paragraph, the experiments are carried out using microvolumes in order to minimise the amount of sample required.

5.2.2 Crystallisation Methods

The first technique, dialysis, is well-known to biochemists. It allows the crystallising agent to be gradually brought into contact with the sample without changing the concentration of the crystallising agent in the solution (figure 5.4). The protein solution and crystallisation solution are separated by a semi-permeable membrane. Salts and other crystallising agents gradually diffuse into the protein-containing compartment. The rate of exchange is controlled by the pore size of the membrane and the concentration gradient. To accommodate microvolumes, this method is often implemented in dialysis buttons with volumes between 5 and 50 μl. This approach is currently not automated and is therefore mainly to optimise crystallisation, *i.e.*, refinement of the crystallisation conditions obtained with large screens.

Alternatively, the sample and the crystallisation agent can be brought together at the start of the experiment and the concentration slowly increases as the test proceeds. A commonly used technique for this is vapour phase diffusion (figure 5.4). The system consists of a reservoir containing between 100 μl and 1 ml of crystallising solution. A micro-drop is formed on a microscope slide or in a suitable cup by depositing a mixture of crystallisation solution and protein solution. This initial mixture dilutes the protein and crystallising agent concentrations by a factor depending on the volumetric proportions of each solution. The slide is then turned upside down so that the drop is suspended above the reservoir. The system is sealed to allow equilibration between the drop and the reservoir. As the volume of the reservoir is much greater than that of the drop and is more concentrated in the crystallising agent, water vapour slowly migrates from the drop into the reservoir. As a result, the drop gradually becomes more concentrated until the concentration of the crystallising agent reaches the same value as in the reservoir. At the same time, the concentration of protein increases and eventually reaches the supersaturated state mentioned above, which favours crystallisation. This technique is known as the hanging or sitting drop technique, depending on the position of the drop containing the protein (figure 5.4). Compared to the hanging drop technique, the sitting drop technique is more automated. It also allows the system to be opened more easily, for example for seeding (see below).

In dialysis and vapour phase techniques, the system gradually evolves towards a state of equilibrium that allows the formation of a seed crystal when the conditions are met.

It may also be interesting to mix the protein solution with a crystallisation solution and allow the system to rest without changing the concentration. In the micro-batch technique, for example, the micro-drop of the mixture is placed either in a cup or in a capillary and covered with oil to prevent, control or limit evaporation (figure 5.4). This technique can also be automated. It is complementary to the vapour phase diffusion approaches.

Finally, direct diffusion across an interface can be induced by a liquid-liquid diffusion approach, for example by superimposing the crystallisation solution and the protein solution in a capillary (figure 5.4). This technique rapidly establishes a continuous gradient of protein concentration in one direction and crystallisation agent in the other, creating a local supersaturation. In some cases, the gel is used instead of the liquid phase. In this way, the concentration gradient is established more slowly and continuously along the capillary. This technique, which is not currently automated, can be very powerful in refining crystallisation conditions.

For initial screening, vapour diffusion is currently the most widely used method. It allows a large number of conditions to be tested with the minimum amount of sample. It is the most widely used method in today's robotic systems and is the preferred method for reproducibility and optimisation of conditions. The other methods are more useful in the optimisation stages or the most difficult crystallisation cases.

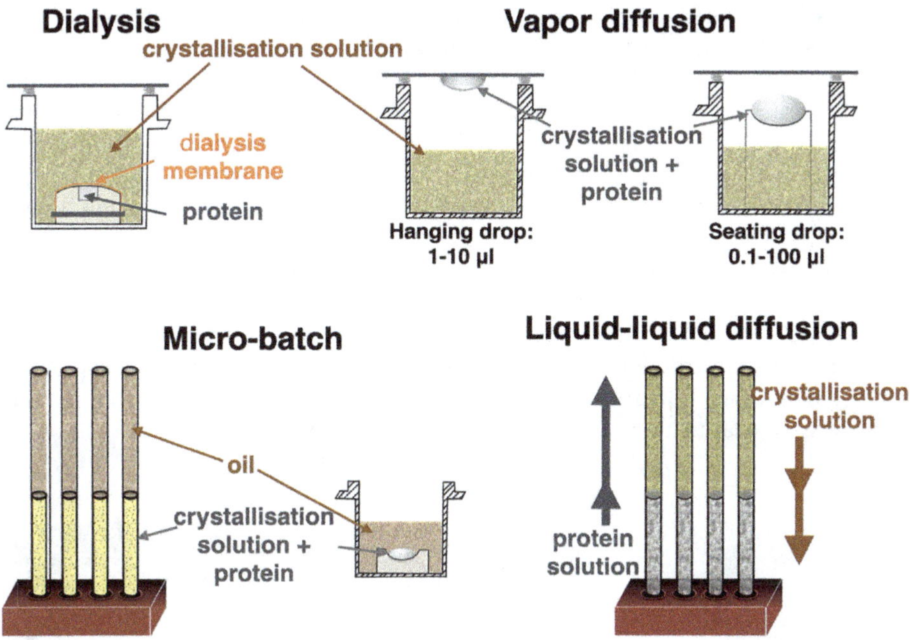

FIG. 5.4 – Some of the common techniques used to crystallise a biological macromolecule.

5.2.3 Crystallisation Approaches

Before a broad screen can be carried out, it is essential to have an idea of the optimal protein concentration for crystallisation. For this purpose, a matrix screen can provide interesting information: for a given protein concentration, different crystallisation agents are tested at different concentrations and different pH values. These preliminary tests can be carried out with very few different conditions and therefore very little protein. If the protein is too concentrated, all the conditions tested will lead to amorphous precipitates; if it is not concentrated enough, the drops will remain clear. This type of experiment allows you to check whether the protein concentration is suitable for commercial screens and to adjust it if necessary.

Once the protein concentration has been adjusted, statistical screens are used to avoid having to test thousands or even tens of thousands of conditions. The first approach, proposed by Charles Carter Junior in 1979, is based on the calculation of an incomplete factorial design: knowing the precipitants, additives and pH values to be tested, all possible associations are calculated, in the form of a huge matrix, and then the associations of the parameters are randomly extracted, with the constraint that two parameters can only be associated once. In 1991, Jarmila Jancarik proposed a matrix sampling approach similar to that of Charles Carter. However, she deliberately biased her approach towards the most common crystallisation conditions found in the literature. Currently, several companies offer commercial screens based on this approach, which can test a large number of conditions. Some companies are now starting to use machine learning algorithms to improve crystallisation screens and optimise crystallisation conditions.

The use of robotics makes it possible to screen a large number of conditions using very small amounts of protein, typically 0.1 μl per experiment. Thus, with the many commercially available screens, several thousand conditions can be tested using a few hundred microlitres of protein. Robots also enable greater reliability and reproducibility of experiments than if these thousands of conditions had to be run manually.

The next step is to analyse the results. A crystallisation experiment is monitored by observing the crystallisation drops, first daily and then weekly. The conditions are scored according to the observations: clear drops, diffuse precipitates, denser or gelatinous precipitates, phase separations, spherulites, needles where the crystal has grown mainly in one dimension, platelets where the crystal has grown mainly in two dimensions, or beautiful crystals with three dimensions (figure 5.5).

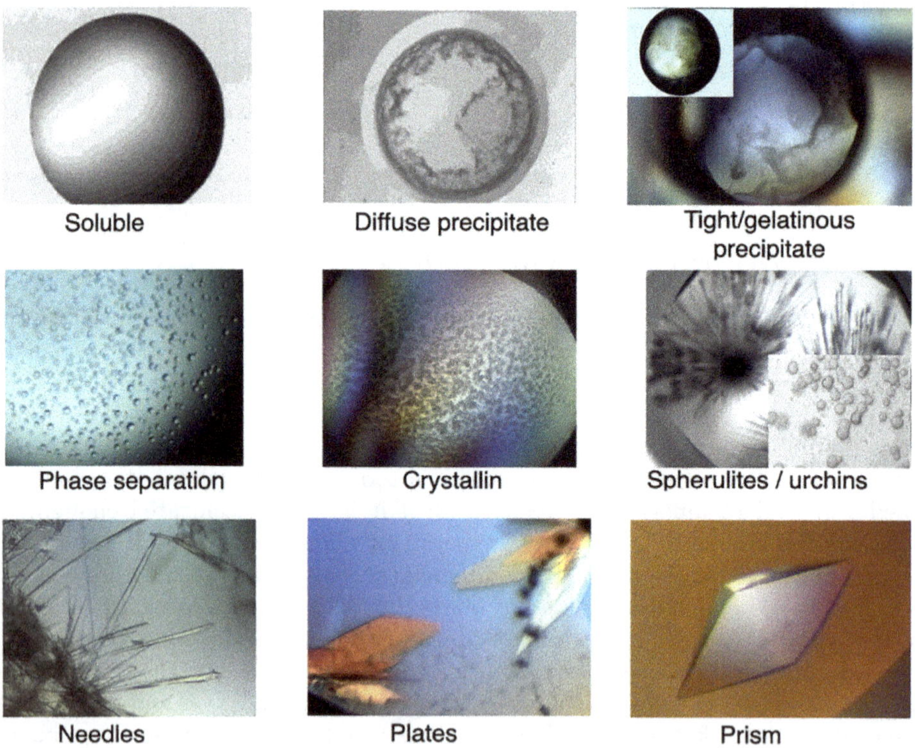

FIG. 5.5 – Examples of crystallisation drops to illustrate the characteristic states that are used to evaluate the results.

5.2.4 Optimisation of Crystallisation Conditions

We first identify whether a crystallising agent, additive and/or pH is more favourable to protein crystallisation. If the initial experiments lead to crystallisation, but there is a need to improve the crystals, it is possible to refine the conditions, for example by varying drop sizes and proportions and using an alternative approach.

We have seen that crystallisation involves two stages: nucleation and growth. The search for optimum conditions can also be applied to the growth phase. For example, it is possible to observe conditions that produce a large number of small crystals that are unsuitable for a diffraction experiment. We can search for different conditions for their growth by seeding from a pre-equilibrated solution at a concentration corresponding to the metastable zone (figure 5.6):

- Macro-seeding, which consists of three steps:
  - o aspiration of the crystal;
  - o washing;
  - o transfer to a pre-equilibrated solution. If the conditions in the pre-equilibrated drop are favourable, the crystal will grow.
- Microseeding, which involves seeding a pre-equilibrated solution with one or more crystal seeds. Two approaches can be used:
  - o crushing;
  - o streak seeding, which involves picking up a crystal or crystal aggregate with, for example, a cat's whisker under a binocular loupe. The cat's whisker is then

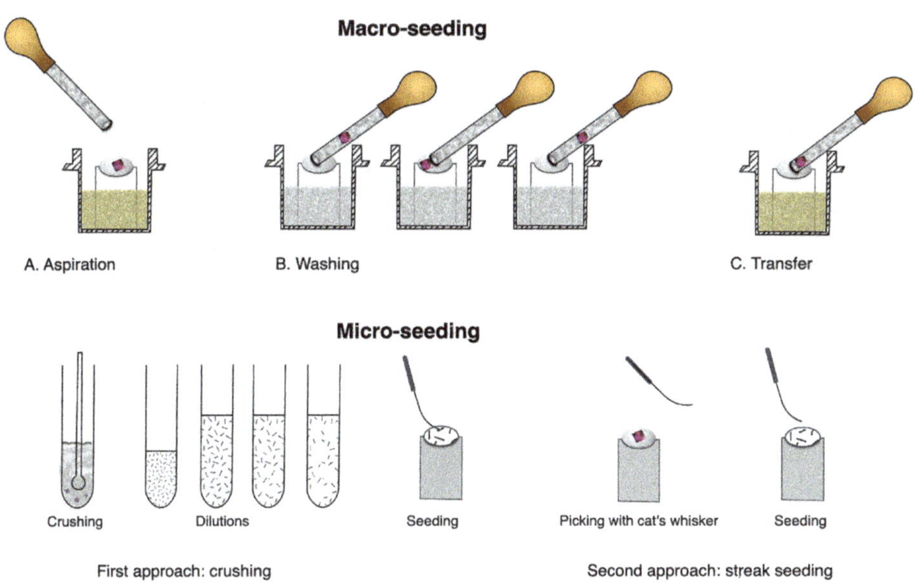

FIG. 5.6 – Crystals optimisation techniques by seeding.

passed through a pre-equilibrated drop. The wheat ear structure of the cat's whisker traps the crystalline germs that are subsequently deposited in a second drop.

In either case, the crystallisation solution may eventually be different from the solution from which the germs originate.

Related videos

Crystallisation: General principles (Cristallisation : principes généraux)
https://youtu.be/CC7m2JM4yu8

Crystallisation: Approaches, platforms (Cristallisation : Approches, plateformes)
https://youtu.be/dgPcErIBCEI

Visit to the laboratory: Crystallisation (Visite du laboratoire : cristallisation)
https://youtu.be/VUxspwIdyXc

Chapter 6

Trip to a Synchrotron

6.1 How to Generate X-Rays?

We have already seen that the penetrating power of X-rays made it possible to photograph the skeleton of the hand of Wilhelm Röntgen's wife (figure 6.1, left). In 1912, Max Laue established the wave nature of X-rays when he observed a diffraction phenomenon in a crystal. He concluded that this was light, an electromagnetic wave with an electric component and a magnetic component (figure 6.1, right) with a very short wavelength.

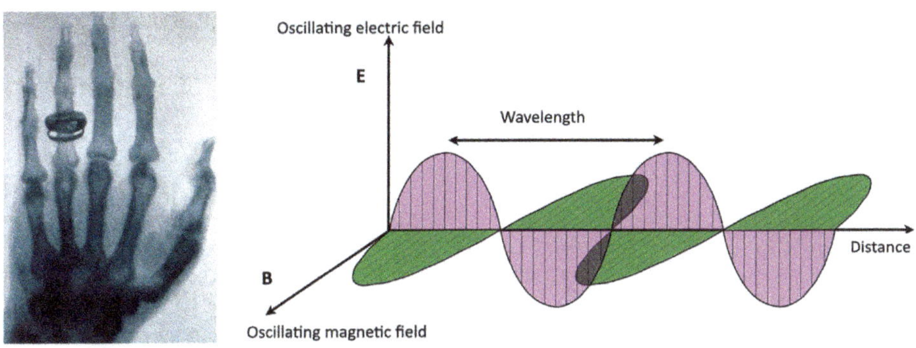

FIG. 6.1 – Left: X-ray radiograph of the hand of Wilhelm Röntgen's wife (https://fr.wikipedia.org/wiki/Wilhelm_Röntgen#/media/Fichier:First_medical_X-ray_by_Wilhelm_Röntgen_of_his_wife_Anna_Bertha_Ludwig's_hand_-_18951222.gif); right: Representation of the electric and magnetic components of an electromagnetic wave.

6.1.1 First X-Ray Generators

At that time, a Coolidge tube was used to produce X-rays (figure 6.2). In this vacuum lamp, a high voltage is applied between a heated metal filament (the cathode) and a metal component (the anode). Electrons are accelerated and projected onto the anode. X-rays are produced at the point of impact of the electron beam.

DOI: 10.1051/978-2-7598-3501-0.c006

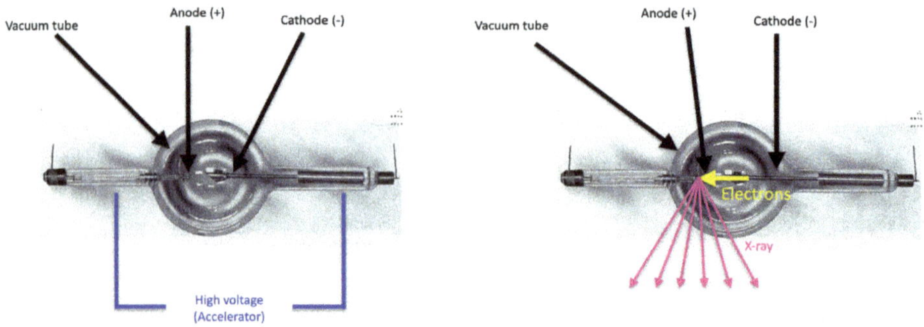

FIG. 6.2 – Production of X-ray from a Coolidge tube (adapted from https://commons.wikimedia.org/wiki/File:Coolidge_xray_tube.jpg).

Two things can happen when the electrons hit the anode. Either the electrons are suddenly slowed down and lose their energy by emitting continuous radiation, the so-called 'brehmstrahlung' – 'decelerating radiation' in English – (figure 6.3), or the electrons are ejected from the inner shells of the atom. The excited atom then relaxes by emitting a fluorescence photon of well-defined energy, characteristic of the electronic energy levels of the atom. The emitted X-rays have a characteristic wavelength that depends on the nature of the anode. Each element in the periodic table is therefore capable of emitting characteristic radiation. However, the elements most commonly used in an X-ray generator for crystallography are copper, molybdenum, silver, or cobalt. Their emission wavelength is on the order of ångstroms, for example, 1.54 Å in the case of copper, which is compatible with atomic structure analysis.

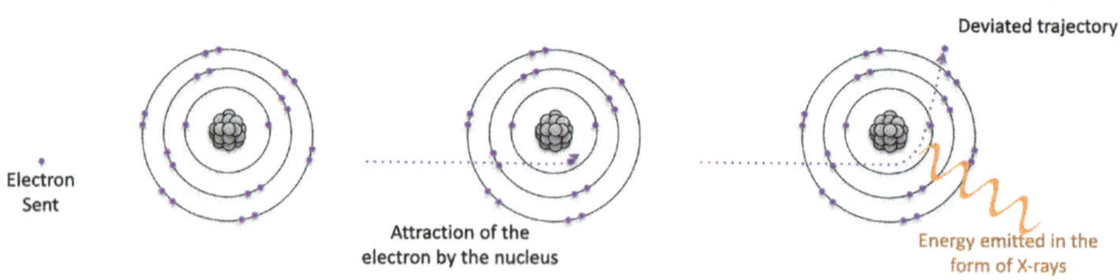

FIG. 6.3 – X-ray emission by brehmsstrahlung.

Based on the general principle of this simple tube, X-ray instruments have undergone major developments in order to improve their performance. The quality of an X-ray beam is usually measured by its brightness, which is a function of the number of photons per second, which depends on the intensity of the beam, the size of the source, and the angle of divergence. For a given flux, X-rays are brighter when they have a smaller diameter and are less divergent. One way to improve the brightness was first to use a heavy metal anode and a higher vacuum, and then to use a rotating anode to limit the heating.

6.1.2 Synchrotron Radiation

Synchrotron radiation is the most common method of producing X-rays used for biological crystallography. Synchrotrons are particle accelerators whose purpose is to produce light (not collisions, as is the case at CERN). An electron gun injects electrons into a first linear accelerator called LINAC and then into a second ring-shaped accelerator called Booster. In the booster, the electrons are accelerated to almost the speed of light. They are then injected into a storage ring, where they continue to rotate with a constant energy. The storage ring consists of a series of magnetic elements (bending magnets, undulators, or wigglers) and radio-frequency cavities that restore the energy lost by the electrons through radiation. As the electrons pass through the magnetic elements, they lose some of their energy in the form of emitted radiation, which is recovered in a laboratory, called a beamline, built tangentially to the storage ring (figure 6.4). Synchrotron light is used in a wide range of disciplines, such as crystallography, but also solid-state physics, chemistry, medicine, and heritage sciences. Therefore, each beamline has its own configuration depending on the type of experiments being performed.

At the beamline, the wavelength of interest is selected and the beam is focused by the optical elements (monochromator and mirrors) located in the optical hutch. Samples are placed in the experimental hutch in an environment appropriate to the type of experiment. The wavelength, size, and intensity of the X-ray beam are controlled from the control room, along with the settings of the sample environment to perform the experiment (figure 6.4).

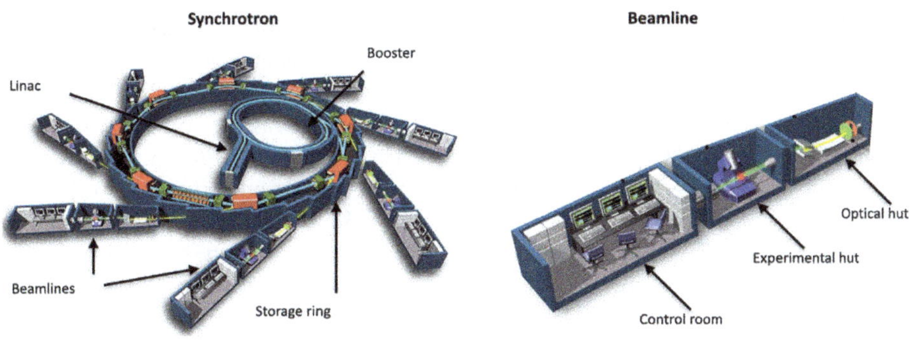

FIG. 6.4 – Picture of the SOLEIL synchrotron on the left, and of a beamline on the right (Source: https://fr.wikipedia.org/wiki/SOLEIL, https://www.synchrotron-soleil.fr/fr/structure-dune-ligne-de-lumiere).

There are about fifteen synchrotron radiation centres in Europe, each with different characteristics. There are two in France: the ESRF (European Synchrotron Radiation Facility) in Grenoble and the SOLEIL synchrotron, on the Saclay plateau, south of Paris.

6.2 Radiation Damage and Crystal Freezing

6.2.1 Radiation Damage

When a crystal is exposed to an intense X-ray beam, it deteriorates rapidly. This is called radiation damage and can be seen as a stain in the area where the beam has passed (figure 6.5).

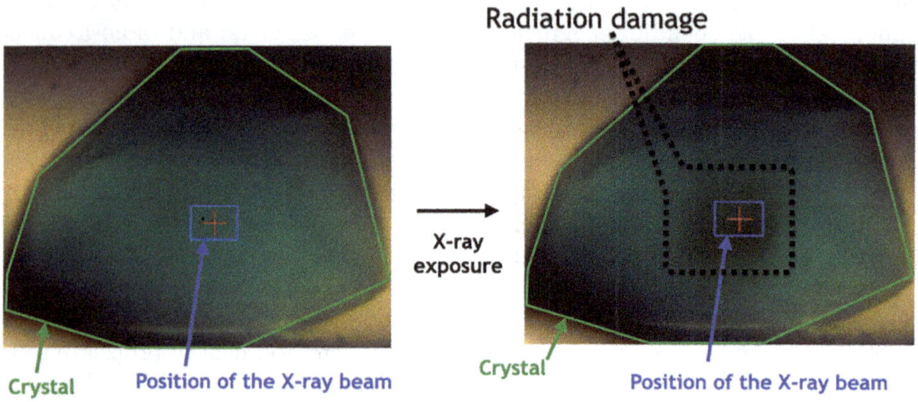

FIG. 6.5 – Radiation damage after exposure of a crystal to an intense X-ray beam.

When a crystal is exposed to an X-ray beam, most of the beam passes through without interaction, and only a small fraction of the beam interacts with the electrons in various ways (see box in chapter 4, **Interaction process between light and atoms**). The interaction we are interested in, the one that causes diffraction, is related to the phenomenon of scattering. However, it is the phenomenon of absorption that can have a detrimental effect on the experiments that we wish to carry out. Absorption transfers a large part of the energy of the X-rays to the crystallised molecules or to the water molecules present in the crystal (see box on the **Production of high energy electrons**). For example, when the oxygen in a water molecule absorbs an X-ray photon, it is ionised and emits a high-energy electron, triggering a cascade of dissociations. This water molecule can then dissociate into two free radicals, hydrogen $H^{\bullet}$ and hydroxyl $OH^{\bullet}$, which are highly reactive chemical species. The formation and propagation of these free radicals can lead to a rapid degradation of the crystal (figure 6.6).

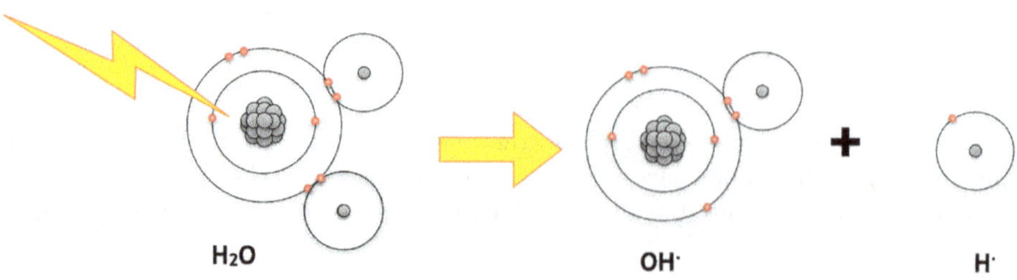

FIG. 6.6 – Formation of $OH^{\bullet}$ and $H^{\bullet}$ free radicals after absorption of an X-ray photon.

Several parameters can determine the occurrence of this phenomenon: the intensity and the energy of the radiation, the type of solvent, the pH value or the temperature. A crystal of biological macromolecules contains on average 50% macromolecules and 50% water molecules. This high solvent content makes macromolecular crystals very sensitive to absorption effects at room temperature, where the solvent is still liquid.

Production of high-energy electrons

The most likely interaction between X-rays and a macromolecular crystal is absorption, which induces the emission of photoelectrons, as seen in chapter 4. The electrons thus ejected from the ionised atoms (primary ionisation) will recover much of the energy originally carried by the X-ray photons. Electrons are much more likely than photons to interact with the surrounding atoms because they are massive and charged particles. These electrons will in turn travel through the crystal over a few micrometres, causing numerous new (secondary) ionisations and excitations of electronic states, gradually transferring their energy and damaging the chemical structure of the molecules in the crystal.

As a result of ionisation, chemical changes can range from disulfide bridge reduction, decarboxylation or the formation of highly reactive radicals, up to the release of gaseous reaction by-products such as H_2 or CH_4. To give an idea, the absorption of a single photon at an energy of 12.4 keV photon (wavelength of 1 Å) very quickly triggers a cascade of events, resulting in several hundred chemical changes over a characteristic distance of 3 μm (figure below). Some ionisations produce coloured zones that make the irradiated volume visible, as shown in figure 6.5.

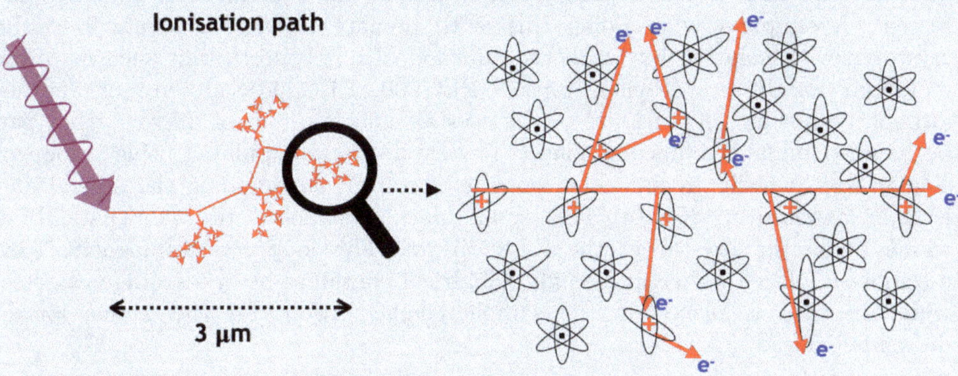

Therefore, the degradation of the crystals is inevitable when they are exposed to X-rays. When the absorbed dose is moderate, only the local chemical structure is affected. As the dose increases, the destabilisation of intra- and intermolecular interactions eventually affects the crystal structure itself. This loss of order is then manifested by a dose-dependent progressive decrease in the maximum diffraction resolution.

6.2.2 Flash Freezing of Crystals

During the acquisition of diffraction data, it is essential to limit the X-ray irradiation in order to obtain a complete data set with minimal damage to the molecules under observation. Low temperatures greatly reduce the incidence of radiation damage, and allow the use of higher doses for a given crystal. Ideally, very low temperatures should be used to limit the propagation of damage. However, it is also important to work at temperatures that are compatible with the tools, equipment and materials of the synchrotron beamlines. The main problem associated with crystal freezing is the water content of the crystal. If the water crystallises as it freezes, it can cause expansion and disorder of the crystal lattice. In addition, the presence of crystalline ice would itself be a source of additional diffraction, preventing the measurement of certain diffraction spots. To avoid this, the crystals are quickly immersed in a liquefied gas at low temperatures. This does not have time for crystalline ice to form and the result is a glassy (amorphous) ice whose density and volume don't change. Another advantage of this glassy form of ice is that it does not diffract. Finally, vitrification strengthens the crystal lattice of the macromolecule, which protects it from radiation damage.

The most commonly used liquid gas is nitrogen, which boils at 77.4 K, or −195.8 °C. Liquid ethane allows the crystals to be cooled more quickly, and in some cases appears to be less damaging to the crystals. The use of a nitrogen gas stream cooled to 100 K is particularly suitable, but it is not always available.

6.2.3 Fast Freezing of Crystals with Liquid Nitrogen

The nitrogen is placed in a suitable, cold-resistant and heat-insulating container. To limit the risk of ice forming in the crystal, the crystal must be 'cryoprotected'. A simple way of doing this is to prepare a solution identical to the crystal growth solution but with the addition of a cryoprotectant such as a low molecular weight polyethylene glycol (PEG200, PEG400), glycerol or another alcohol, or a sugar. As the size of the crystals is a few tens of microns, they can be handled under a binocular loupe. The crystal is manipulated using a loop of similar size in order to limit the amount of solvent surrounding the crystal and is later exposed to X-rays. This small loop is attached to a magnetic base which holds the crystal in the X-ray beam. The loop/crystal/magnetic base assembly is placed either in a small cylindrical container or in a storage cassette, which in turn is placed in the sample changer robot for the acquisition of diffraction data.

To freeze a crystal, we proceed as follows (figure 6.7):

1/ we prepare a drop of crystallisation solution mixed with cryoprotectant, place it under the binocular loupe and focus it;
2/ we position the drop of cryoprotectant near the loop;
3/ we 'fish' the crystal;
4/ using the loop, we gently place the crystal in the cryoprotectant solution for a few seconds;
5/ we retrieve the cryoprotected crystal with the same loop;
6/ we immediately plunge the loop with the crystal in the liquid nitrogen solution containing the storage cartridge.

Once the crystal is frozen, the storage cassette is placed in a transport container to take it to the synchrotron. This example shows an approach that has the advantage of being simple and fast in the first instance.

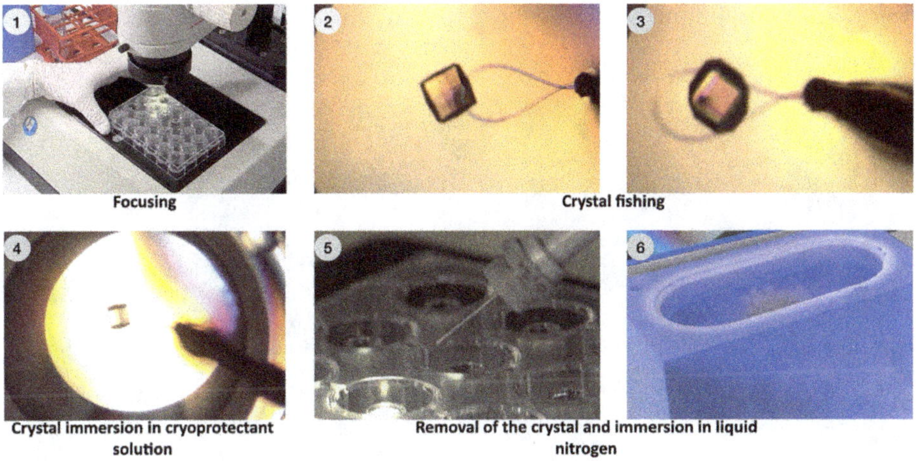

FIG. 6.7 – Steps involved in cryoprotecting the crystal.

6.3 Experimental Hutch: The Crystal Environment

6.3.1 Example of PROXIMA-1 Beamline

To understand what happens in a beamline, we have chosen the example of the PROXIMA-1 beamline at the SOLEIL synchrotron (figure 6.8). Data acquisition takes place in the experimental hutch. As the X-ray flux is very intense, this cabin is closed during the entire data acquisition to ensure the safety of the users, but all the devices can be remotely operated from the control room. The cassettes containing the crystals are stored in a container of liquid nitrogen, which can contain between 3 and 9 cassettes, *i.e.*, several dozen crystals. Each crystal is positioned under the X-ray beam using a robot whose mechanical arm picks up a crystal and places it on a device called a goniometer, which allows the crystal to be centred and aligned in the X-ray beam. A stream of nitrogen gas cooled to 100 K is continuously applied to keep the crystal at a low temperature.

When an X-ray beam illuminates a crystal, most of the beam passes through the crystal without being diffracted. This very intense direct beam would damage the detector and must be blocked by a small lead well, called a beam stop. Above the crystal, a fluorometer can measure the X-ray fluorescence signal to provide information about the elemental composition of the crystal. The detector, which records the diffraction data, is on the opposite side of the X-ray beam. The distance between the detector and the crystal has a direct effect on the maximum 2θ angle (see box **Resolution limit and wavelength**, chapter 4) and therefore on the resolution of the data set. It also affects the distance between the diffraction spots. For each sample, this distance is determined by the lattice parameters of the crystal and its diffraction power, *i.e.* the resolution to which the crystal diffracts X-rays (figure 6.8).

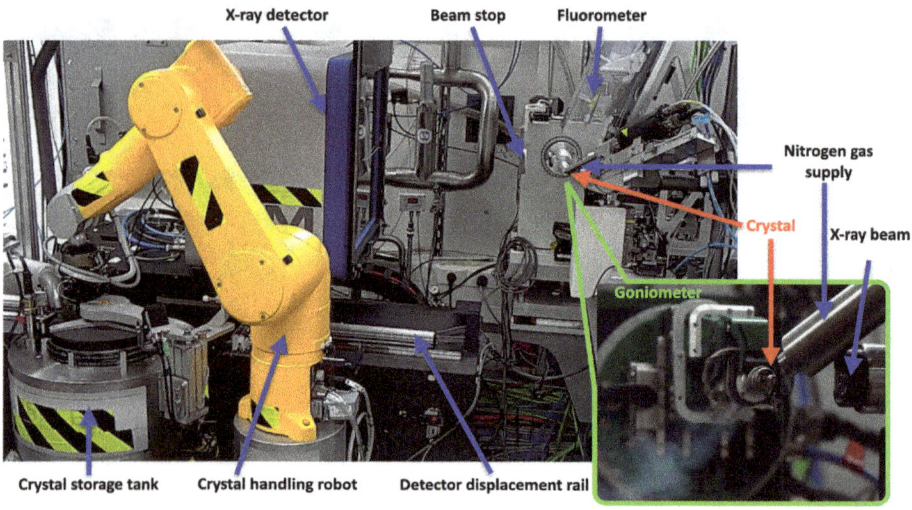

FIG. 6.8 – Overview of the crystal environment in the PROXIMA-1 beamline experimental hut.

Since the expected result is the three-dimensional structure of the macromolecule under investigation, we need to collect the diffraction data in all three dimensions. To do this, we rotate the crystal while collecting the data and simultaneously record a series of images, each corresponding to a small part of the rotation. Each diffracted ray hits a sensitive layer of the detector. The sum of the rays diffracted in one direction gives a diffraction pattern. Each diffraction image therefore contains a set of diffraction spots. A diffraction data set consists of a series of images which, when put together, allow the reconstruction of a three-dimensional lattice, the reciprocal lattice.

6.3.2 The Control Room

In the control room, 4 screens allow monitoring (figure 6.9):

- the movements and correct operation of the crystal storage and passage robot;
- the whole crystal environment;
- the centring of the crystals in the X-ray beam, and the control of the data acquisition;
- the observation of the diffraction patterns of the crystals.

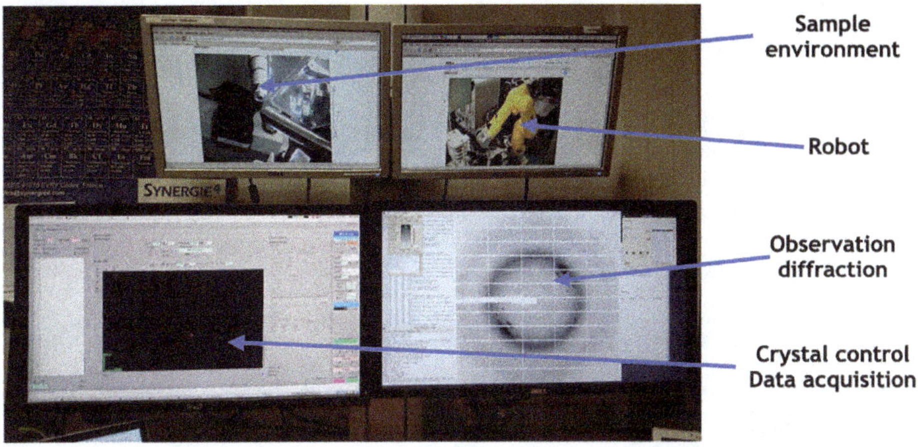

FIG. 6.9 – Control of sample environment, robot operation, crystal visualisation and diffraction.

A dedicated interface can be used to control the robot that positions a crystal on the goniometer. After logging on to the interface using the identifiers associated with the laboratory, the crystal is selected from the various positions in the storage cassette displayed on the left of the screen. The robot transfers the selected crystal to the goniometer. Once positioned, the crystal must be aligned so that it remains in the beam while rotating on its axis to collect a complete data set (figure 6.10). As the X-ray beam is very small (on the order of a few tens of microns in diameter), the crystal must be centred within this small volume.

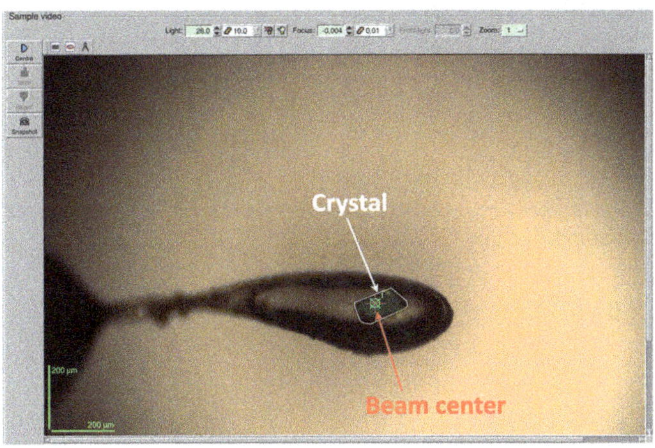

FIG. 6.10 – Visualisation of the crystal positioned on the goniometer. The control panel on the top left allows to illuminate, magnify the visualisation and centre the crystal on the beam.

Once the crystal is centred, we can observe its diffraction properties in the X-ray beam by exposing it in different orientations.

6.3.3 Diffraction Data

Diffraction occurs when a crystal is illuminated with X-rays. This can only happen because the wavelength of the X-rays is the same order of magnitude as the inter-atomic distance. Each spot in the diffraction pattern corresponds to a diffracted ray. It is characterised by:

- a position described by three integers: h, k and l, called the Miller indices,
- an amplitude $|F|$,
- a phase α.

The amplitude is proportional to the square root of the intensity of the spot. In a diffraction experiment, the coordinates of the spots and their intensities are measured from the diffraction images and the amplitudes $|F|$ and the indices hkl are deduced (see §7.1). However, the phase cannot be measured directly. The whole mathematical problem of crystallography arises from the so-called phase problem that has to be solved (figure 6.11).

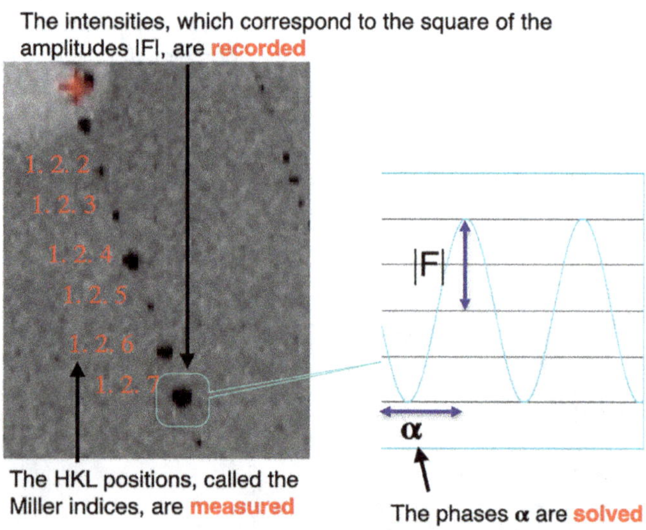

FIG. 6.11 – Characteristics associated with each diffraction spot.

Related videos

How to generate X-rays (Comment générer des rayons X)
https://youtu.be/8SLr51Iblk8

Freezing crystals (La congélation des cristaux)
https://youtu.be/rWG2Rqx2BEc

The crystal environment in the hutch (L'environnement du cristal dans la cabane)
https://youtu.be/l69Ebuj0SaU

Chapter 7

Acquisition, Processing and Analysis of Diffraction Data

7.1 Collection Strategy

In theory, it should be necessary to collect 360° diffraction images for a complete data set. We will see later that this is generally not the case.

7.1.1 Crystal Characterisation

We now need to define the parameters for the first data collection:

- The energy (in electron volts) at which we want to collect the data, or the wavelength (in ångstroms), which gives the same information as the relationship between wavelength and photon energy is:

$$\lambda(\text{Å}) = 12.4/E\ (\text{keV})$$

- The maximum resolution (in ångstroms), *i.e.* the distance of the detector from the crystal (in millimetres), which also gives the same information and depends on the diameter of the detector (see box on the **Resolution limit and wavelength** in chapter 4).

To determine these parameters, we usually record two test images at 90° to each other, *i.e.* by rotating the crystal by 90° on itself. From this stage on, we work in the space of the diffraction data, the reciprocal space (see §3.4). The test images appear in the control window one after the other as they are collected. If the overall intensity level varies from image to image, this may be due to the morphology of the crystal and the amount of frozen liquid around the crystal. It may also be due to a simple problem with the centring of the crystal in the beam, which can be corrected. The diffraction quality may also vary depending on the illuminated part of the crystal. It may therefore be worthwhile to test several regions.

From a single diffraction image, we obtain information that allows us to define the data acquisition parameters (figure 7.1):

DOI: 10.1051/978-2-7598-3501-0.c007

- The white area in the centre of the image is the shadow of the beamstop, blocking the direct beam.
- The resolution depends on the wavelength used and the maximum theta angle (see box in chapter 4 on **Resolution limit and wavelength**). Since the size of the detector is fixed, the maximum theta angle varies with the distance between the detector and the crystal (see figure 4.8). Therefore, the resolution at the edge of the detector depends on this distance. The information about the beam wavelength and crystal-to-detector distance is entered into the software, which calculates the resolution and displays the value along the circles centred on the direct beam. The image in figure 6.11 shows diffraction spots down to about 2.0 Å, which is the resolution limit for this crystal.
- Around 3.8 Å there is a ring of diffuse scattering which increases the background noise. This ring is due to the scattering from the water in the crystal and from the solution surrounding the crystal. To reduce the solvent scattering, we try to confine the solvent around the crystal by fishing it with a loop of similar dimensions.
- The distance between the diffraction spots varies with the direction. The reciprocal space is observed and there is an inverse relationship between the spacing of the spots and the size of the cell (see §3.4). The closer the spots are to each other, the larger the lattice parameter in that direction.

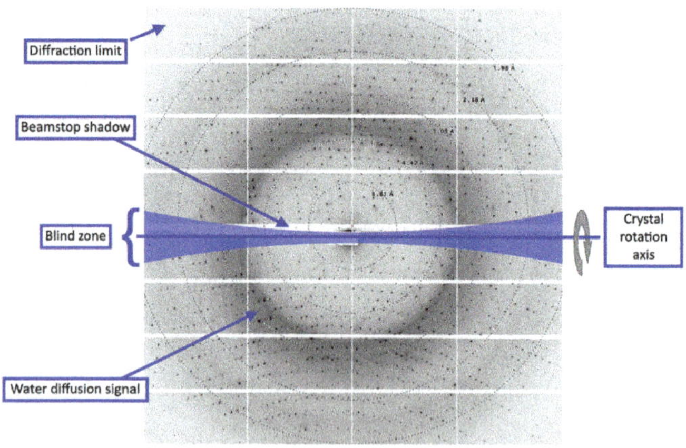

FIG. 7.1 – Information in the diffraction image.

In figure 7.1, we can also see that the intensity of the spots is higher near the centre of the diffraction pattern, and decreases until it disappears as we move away from the centre (see the box in chapter 4 on the **Atomic scattering factor**). The signal extinction defines the resolution limit of the data, *i.e.*, the smallest details that we can observe. Knowing the resolution limit of the crystal allows us to optimise the distance between the detector and the crystal. Around the rotation axis of the crystal, the Bragg spots are too elongated and cannot be used. This region defines the so-called blind zone (figure 7.1).

The so-called 'Laue group' consists of symmetry operations that leave the diffraction pattern of a crystal unchanged. These symmetries induce equivalences in the reciprocal lattice that reduce the minimum amount of data needed to describe the entire content of the lattice. We can also calculate the lattice parameters from just a few images, using software that reads the images and locates the position of the diffraction spots. The software calculates the position of the spots relative to each other and to the position of the direct beam. Knowing these positions makes it possible to estimate the reciprocal lattice and search for its symmetries. This allows us to calculate the parameters of the direct lattice and its primitive cell (figure 7.2). From these parameters, the software adds constraints corresponding to the different Bravais lattices compatible with the primitive cell. For each lattice, it calculates the degree of agreement between the observed and the calculated lattice. If the agreement is above a certain threshold, the lattice is considered to be incompatible with the symmetry of the crystal under investigation. For example, in the table shown in figure 7.2, the first column lists the different Bravais groups briefly tested, the values of the parameters a, b, c and α, β, γ, and the penalties (quality of fit) associated with the different possible cells. These penalties are very small up to a certain level of symmetry constraint. Then there is a significant jump in the penalty, which excludes higher-order symmetries. This initial calculation of the crystal cell is the basis for determining the orientation of the crystal.

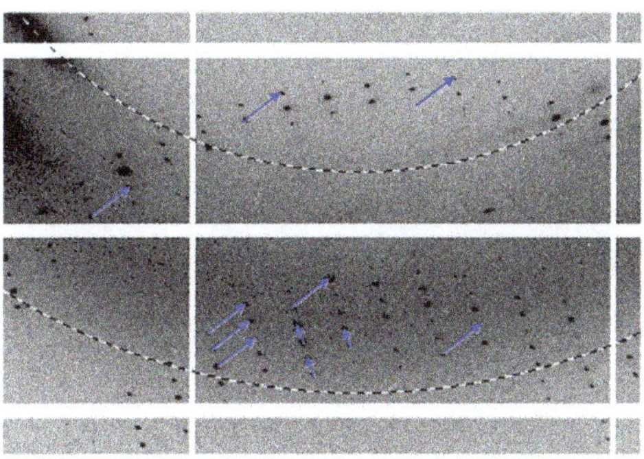

BRAVAIS-LATTICE	QUALITY OF FIT	UNIT CELL CONSTANTS (ANGSTROEM & DEGREES) a	b	c	alpha	beta	gamma
aP	0.0	56.4	83.5	106.3	90.0	90.0	90.0
aP	0.2	56.4	83.5	106.3	90.0	90.0	90.0
mP	0.6	56.4	83.5	106.3	90.0	90.0	90.0
mP	0.7	56.4	106.3	83.5	90.0	90.0	90.0
mP	0.8	83.5	56.4	106.3	90.0	90.0	90.0
oP	0.9	56.4	83.5	106.3	90.0	90.0	90.0
mC	249.7	56.4	219.9	83.5	90.0	90.0	75.2
mC	249.8	56.4	176.2	106.3	90.0	90.0	71.3
mC	250.0	219.9	56.4	83.5	90.0	90.0	75.2
oC	250.2	56.4	219.9	83.5	90.0	90.0	104.8
mC	250.3	176.2	56.4	106.3	90.0	90.0	71.3
oC	250.5	56.4	176.2	106.3	90.0	90.0	108.7
mC	499.4	176.2	56.4	135.1	90.0	125.8	71.3
mC	547.3	83.5	228.3	56.4	90.0	90.0	68.6
mC	547.5	228.3	83.5	56.4	90.0	90.0	68.6
oC	547.8	83.5	228.3	56.4	90.0	90.0	111.4
mC	594.6	100.8	100.8	106.3	90.0	90.0	111.9
mC	594.8	100.8	100.8	106.3	90.0	90.0	68.1
oC	595.2	100.8	100.8	106.3	90.0	90.0	68.1
tP	595.4	56.4	83.5	106.3	90.0	90.0	90.0
oF	624.2	56.4	176.2	219.9	85.3	104.8	108.7
mC	678.7	135.1	135.1	56.4	90.0	90.0	103.7

FIG. 7.2 – Part of a diffraction pattern showing examples of distances between diffraction spots (blue arrows on the left) to determine the cell parameters a, b, c, α, β, γ, and the most likely Bravais lattice (right).

The aim is to quickly collect a complete data set in order to minimise the exposure time of the crystal in the beam. The data acquisition strategy therefore consists in:

- Optimising the initial orientation of the crystal according to the calculated cell.
- Choosing the minimum number of images to be collected according to the possible symmetries for the assumed Bravais group.

7.1.2 The Anomalous Signal

In some cases, in addition to the macromolecule of interest, the crystals contain one or more atoms that can be used to calculate the phases associated with each diffracted intensity. Some of these atoms, such as the selenomethionines discussed in chapter 2, modulate the intensity of the diffraction spots at specific wavelengths. This is known as the anomalous signal, and its significance is discussed in detail in chapter 11. The fluorescence detector above the goniometer (figure 6.8) detects the presence of an anomalous scatterer in the crystal. To do this, we perform an energy scan, in which the energy (and therefore the wavelength) of the X-ray beam is varied stepwise around the absorption edge of the element of interest. In the case of zinc, for example, a sharp change in fluorescence signal intensity can be observed between 9.66 and 9.67 keV (figure 7.3). To maximise the intensity differences, we record data at wavelengths corresponding to the:

- fluorescence peak,
- absorption edge,
- remote wavelength.

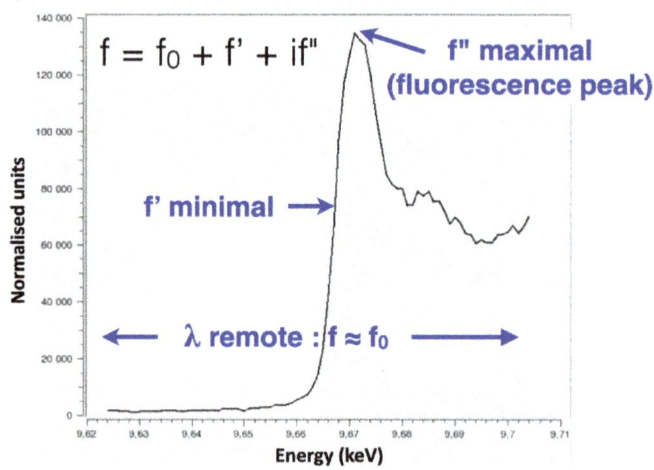

FIG. 7.3 – Fluorescence spectrum of zinc. Intensity of fluorescence radiation as a function of incident photon energy, near the absorption edge of Zn.

If we collect a single data set at the wavelength corresponding to the fluorescence peak, this is called SAD data, for Single Anomalous Dispersion. If we collect data sets at several different wavelengths, this is called MAD data, for Multiple Anomalous Dispersion. It takes between 2 and 10 min to collect a complete data set, depending on the exposure time and the number of images required.

7.2 Diffraction Data Processing

During data acquisition, images are recorded using a specific detector (see box for a **Brief history of two-dimensional detectors**). Each image recorded during data acquisition corresponds to about one-tenth of a degree of rotation of the crystal. The purpose of the data processing is to convert the diffraction images into a set of intensities listed in a file. For each recorded reflection, this includes a position, the Miller indices *h*, *k*, *l*, an intensity and an estimate of the measurement error (figure 7.4), known as the sigma of the intensity.

The data processing steps include the cell parameter determination or indexing step discussed above. Knowing the cell parameters allows us to determine the orientation of the crystal during acquisition and to assign *hkl* indices to each diffraction spot. Once the cell parameters have been refined, the *hkl* positions of all the Bragg reflections can be calculated and compared with the recorded data. The software that processes the data requires a lot of information that can be defined in a text file or in the header of the image files. This includes the physical parameters of the acquisition, such as the wavelength, the distance between the detector and the crystal, and the parameters that define the detector.

Brief history of two-dimensional detectors

In 1895, W. Röntgen discovered X-rays thanks to a fluorescent screen placed a few metres away from his X-ray tube. This was the first X-ray detector! X-rays excite atoms, which de-excite and emit low-energy light. Unfortunately, the fluorescence signal disappeared as soon as the X-rays were switched off. Röntgen then used a very simple detector, which was used until the 1980s: photographic film. These films contain silver halide crystals (*e.g.* $Br^{-}Ag^{+}$). An X-ray photon ionises the Br and the ejected electron reduces a few silver atoms which, under the action of a developer, become small grains of metallic silver of a few tens of microns in size, clearly visible. The films are sensitive and spatially well resolved ($\approx$20 μm), but they saturate quickly. They are linear only over two or three orders of magnitude in intensity: their dynamic range is small.

In the late 1980s, new detectors called 'imaging plates' appeared. These are flexible plastic plates coated with a powder of phosphorescent microcrystals, typically barium fluorobromide doped with europium. When exposed to X-rays, the europium ions release an electron that is trapped in a gap between bromine atoms, forming a long-lived 'colour centre'. During reading, the plate is scanned by a laser beam, which de-excites the centre, and emits a photon of light. The efficiency of this detector is much better than that of silver film, its resolution is lower (100 μm), but its dynamic range is four or five orders of magnitude greater. Image plates quickly replaced film... and were soon replaced by digital detectors in the 1990s.

The first digital detectors were made of semiconductor materials, very often Si or Ge. They all consist of a two-dimensional array of photosensitive sensors, whose task is to recover the electrons or holes formed in the semiconductor by the absorption of a photon. In CCD (charge-coupled device) detectors, detection is usually indirect: a scintillator layer converts the X-rays into visible photons that excite the semiconductor. This type of detector is much more efficient than image plates, and its readout is extremely fast, making it very popular.

Despite other experimental devices, such as wire chambers inspired by the work of Georges Charpak, the revolution in detectors came at the end of the 1990s, with the advent of 'hybrid pixel' detectors, a direct spinoff of the CERN (European Organisation for Nuclear Research) detectors. In this technology, each pixel is a sensor containing an amplifier and a discriminator. Unlike previous detectors, they do not integrate the signal but count the photons. They also have no noise. With the advances in microelectronics, the pixel size can now be less than 100 μm.

The images are processed in successive steps (figure 7.4):

- The first step is to evaluate the background noise of the images. This background is caused by the continuous scattering of X-rays that have no constructive interference. The average number of counts in each pixel and the average variation (standard deviation) of these values for neighbouring pixels are determined from a few images in the dataset. These values are used to estimate the background noise in the image set. After subtracting the average background surrounding a group of pixels, the diffraction signal can be distinguished for pixel values N times above the standard deviation.

- The second step is to search for diffraction spots. To perform the indexing, it is usually sufficient to look for these diffraction spots on two sets of images recorded by rotating the crystal by 90°. A set of contiguous pixels with a cut-off value larger than the average of the background is considered to be a diffraction spot. At the end of this step, we have a list of potential diffraction spots with an intensity (sum of the counts for the contiguous pixels) and a position (the centroid for X and Y on the image and for the rotation angle of the crystal) for each one.

- The third step is automatic indexing. It consists in determining the possible Bravais lattice types and the cell parameters. This step was usually carried out once before the data acquisition in order to define the strategy (see chapter 6). However, a large number of images available increases the reliability of the determination of the cell parameters. These parameters are essential for indexing the set of spots, *i.e.* assigning the Miller indices *hkl* to each reflection. At this point, we define the cell parameters and the orientation parameters of the crystal. We also formulate a hypothesis about the symmetry and the space group, which then needs to be verified. On the basis of this information, the software assigns an *hkl* position to each new spot detected in an image.

- The fourth step is the integration. This involves an estimate of the diffraction intensity and the background associated with each diffraction spot. A diffraction spot is not a single point, but occupies a specific three-dimensional volume. As fine oscillations are recorded for each image, most diffraction spots are recorded in several successive images. To obtain the total intensity at a position *hkl* it is therefore necessary to sum these partial intensities. To do this, the average shape of a diffraction spot must be accurately defined. This is called the three-dimensional profile. Integration is the most time-consuming step since the software analyses the images of the entire dataset one by one and sums the intensities of the measured pixels around the expected *hkl* positions for each diffraction spot.

- The fifth step is called scaling. Once all the intensities have been summed over all the images, this very last step consists of applying correction parameters to take into account experimental errors and parameters related to the configuration of the experiment. This final stage of corrections allows us to assess the quality of the collection and to validate the chosen symmetry hypothesis. We test the different symmetry possibilities for a given cell geometry and select the one with the best internal agreement (see below).

- The final step is the choice of the space group. At this point, we have all the intensities that make up the reciprocal lattice of the crystal, and we have assigned a Bravais lattice. We have made a first hypothesis about the space group without taking into account the presence of possible screwed (or helical) axes, it is called the point group. To complete the space group assumption, we use a property of screw axes, namely that they induce characteristic reflection conditions along the reciprocal lattice axes. For example, axis 2_1 can be identified by the reflection conditions of every other reflection in the corresponding direction. Thus, from a $P222$ point group, reflection conditions of the type $l = 2n$ along the axis $[00l]$ are characteristic of the space group $P222_1$, whereas reflection conditions of the type $h = 2n$ along the axis $[h00]$ and $k = 2n$ along the axis $[0k0]$ are characteristic of the space group $P2_12_12$. However, it should be noted that some space groups remain indistinguishable at this stage, they are enantiomorphic. For example, in the case of a point group $P4$, the 4 space groups $P4$, $P4_1$, $P4_2$ or $P4_3$ can be specified according to the possible characteristics of the helical axis. In the space group $P4$, there are no reflection conditions, in the group $P4_2$ they are $l = 2n$ along the axis $[00l]$, and in the group $P4_1$ they are $l = 4n$ along the axis $[00l]$ as for the group $P4_3$. The space groups $P4_1$ and $P4_3$ are therefore enantiomorphic, they can only be distinguished once the structure has been solved.

At the end of the data processing, we obtain a file with the intensities of the *hkl* reflections, and their statistics, in order to evaluate their quality.

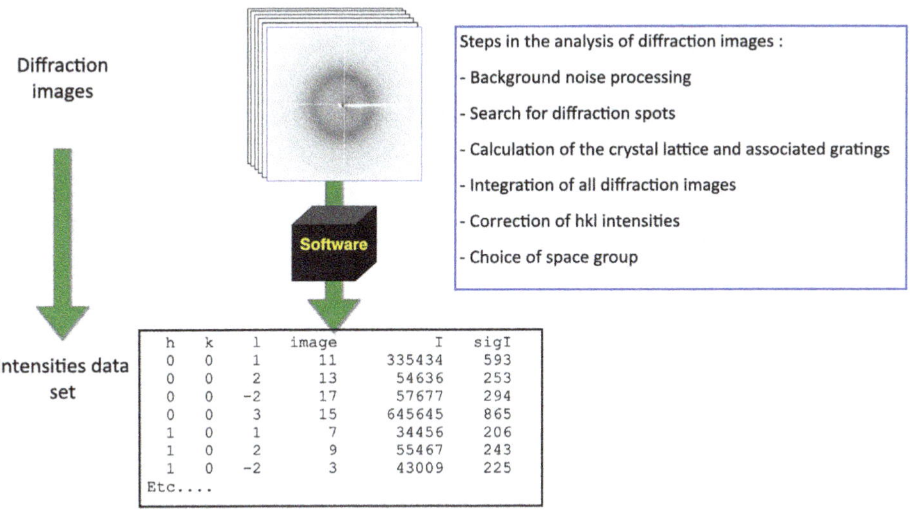

h	k	l	image	I	sigI
0	0	1	11	335434	593
0	0	2	13	54636	253
0	0	-2	17	57677	294
0	0	3	15	645645	865
1	0	1	7	34456	206
1	0	2	9	55467	243
1	0	-2	3	43009	225
Etc....					

FIG. 7.4 – Summary of data processing objectives and steps.

7.3 Analysis of Diffraction Data

On completion of the data integration, we have a set of 'raw' intensity measurements, each associated with *hkl* Miller indices. At this stage, it is necessary to evaluate several possible symmetry assumptions. To select the symmetry model that best explains the internal symmetry relationships between the intensity measurements, we compare the intensities to each other to identify groups of reflections that are equivalent, and we test our hypothesis using statistical criteria.

7.3.1 Friedel Law

Friedel's Law (to be discussed in more detail in §11.2) tells us that:

$$I(hkl) = I(\overline{hkl})$$

The consequence of this law is a centre of inversion in the reciprocal space. Therefore, we usually only need to collect 180°, *i.e.* half of the reciprocal space, to obtain a complete data set. Friedel's law is almost always respected, except for non-centrosymmetric crystals in the presence of an anomalous scatterer (see §11.2).

7.3.2 The Choice of the Laue Group

From Friedel's law, we already know that *hkl* reflections are equivalent to $\overline{hkl}$ reflections, except in the vicinity of an absorption edge where this rule is not satisfied. The symmetries present in the crystal induce additional equivalences between groups of reflections. These sets of equivalences are classified into groups called Laue groups.

7.3.3 Assessment of Data Quality

The statistical criteria for evaluating the data are based on both measures of mean differences (denoted by *R*...), and correlation coefficients (abbreviated as *CC*). By convention, we calculate what we call an internal agreement factor, or *R* factor, defined as:

$$R(\%) = 100 \times \frac{\sum_i^n \left| |F_i|^2 - \langle |F_i|^2 \rangle \right|}{\sum_i^n |F_i|^2} \tag{7.1}$$

Each measured *hkl* reflection intensity is compared to the average value of all reflections at the *hkl* position or an equivalent position, defined by Friedel's law and by the symmetry operations of the Laue group. If the Laue group is incorrectly estimated, the reflections will not be equivalent, so the deviation between the intensities and the average of the intensities will be high, and the *R* factor will also be high. Typically, in biological crystallography, a good quality *R* factor is below 10% for the strongest reflections (in the first rows of the statistics table in figure 7.2).

RESOLUTION LIMIT	NUMBER OF REFLECTIONS OBSERVED	UNIQUE	POSSIBLE	COMPLETENESS OF DATA	R-FACTOR observed	R-FACTOR expected	COMPARED	I/SIGMA	R-meas	CC(1/2)	Anomal Corr	SigAno	Nano
4.94	12488	1703	1706	99.8%	3.0%	2.9%	12486	57.94	3.2%	99.9*	-9	0.842	1548
3.51	21931	3001	3001	100.0%	3.2%	3.1%	21930	53.99	3.4%	99.9*	-4	0.855	2797
2.87	28989	3873	3873	100.0%	4.6%	4.2%	28988	37.69	4.9%	99.9*	-3	0.857	3598
2.48	34331	4582	4582	100.0%	8.5%	8.0%	34330	21.13	9.1%	99.7*	-1	0.848	4240
2.22	38879	5190	5190	100.0%	13.3%	13.2%	38878	13.95	14.3%	99.3*	-3	0.792	4761
2.03	41924	5663	5664	100.0%	20.8%	20.9%	41923	9.04	22.3%	98.4*	-1	0.787	5120
1.88	45568	6220	6220	100.0%	37.1%	36.8%	45568	4.73	39.9%	95.2*	-8	0.769	5596
1.76	48922	6654	6653	100.0%	74.1%	80.0%	48922	2.15	79.7%	81.5*	-6	0.698	5916
1.66	48747	6997	7070	99.0%	130.4%	153.8%	48697	0.99	140.7%	53.9*	-2	0.621	6117
total	321780	43884	43959	99.8%	6.4%	6.3%	321722	15.44	6.9%	99.9*	-4	0.766	39693

NUMBER OF REFLECTIONS IN SELECTED SUBSET OF IMAGES	325027
NUMBER OF REJECTED MISFITS	3201
NUMBER OF SYSTEMATIC ABSENT REFLECTIONS	46
NUMBER OF ACCEPTED OBSERVATIONS	321780
NUMBER OF UNIQUE ACCEPTED REFLECTIONS	43884

FIG. 7.5 – Example of a quality evaluation statistics table for a diffraction data set.

The use of correlation coefficients is also based on the assumption of equivalent reflections. The trick here is to divide the measured reflections into two randomly distributed groups. In each group, the equivalent reflections are averaged and then the correlation between the assumed equivalent intensities of these two groups is calculated. This is what we call *CC*(1/2), or *CCano*(1/2) if we are analysing anomalous differences.

In the file containing the statistics related to the data processing, the information is classified by resolution range, in order to determine the limit of the usefulness of the processed data. The information that can be found generally includes (figure 7.5):

- The number of measured (observed) reflections, *i.e.* the total number of reflections collected, without taking into account the equivalent reflections.
- The number of unique reflections, *i.e.* the number of observed reflections after removing the equivalent reflections.
- The number of possible reflections, *i.e.* the maximum number of reflections at a given resolution, these are the unique reflections. They can be calculated from the cell parameters and the selected symmetry.
- Completeness, which can be obtained directly from the ratio of the number of unique reflections measured to the number of possible reflections.
- Multiplicity, sometimes called redundancy, *i.e.* how many times a reflection has been measured on average.
- The *R* factor, which can have different values depending on the reflections used in its calculation (observed, expected, *Rsym*, *Rmerge*, ...), increases with the resolution. This is due to the fact that intensities are smaller at higher resolutions, and therefore more difficult to measure accurately. If one makes a mistake in the choice of the Laue group (see §3.2), the *R* values will be high at low resolution. The *Rmeas* factor (for measured) includes a correction factor to account for multiplicity and avoid underestimating measurement errors.
- The I/sigma ratio, *i.e.* the signal-to-noise ratio. The standard deviation, or sigma, is associated with each measurement. It estimates the background intensity associated with the measurement of each reflection, taking into account the average background noise around the diffraction spots measured during the integration, and then correcting it by the standard deviation of equivalent reflections in the same resolution range. The average value of the

I/sigma ratio is important in assessing the data quality. Typically, in the absence of crystal degradation, the more equivalent reflections (*i.e.* the more redundancy), the higher the final I/sigma should be, reflecting better data reliability.
- The correlation coefficient $CC(1/2)$ defined above. The asterisk marks the values that are statistically significant because they are calculated from a sufficient number of reflections.

All these parameters, *I/sigma*, $CC(1/2)$, R,... allow one to evaluate the overall quality of the measured data and to limit the resolution at which they can be used. The cut-off criteria have evolved over time, and vary from one crystallographer to another. Currently, the accepted criteria are to limit the maximum resolution of the data when the I/sigma ratio becomes less than 1, and/or the $CC(1/2)$ value becomes less than 50%. The R is used to evaluate the quality of the measurements and the choice of the Laue group.

Once the data are scaled, the intensities I must be converted into amplitudes F.

Thus, after conversion, a file of the form (h, k, l, I, *sigI*) becomes (h, k, l, F, *sigF*); or becomes (h, k, l, F^+, $sigF^+$, F^-, $sigF^-$) if an anomalous signal is suspected, in order to separate Friedel pairs (of intensity F^+, for a reflection at position *hkl* and F^- for a reflection at position $-h$ $-k$ $-l$). A final column, called *FreeR_flag*, allows reflections to be marked at random. These reflections are used to calculate a special *R-factor* during the model refinement stage, the *R-free* (see §11.1).

The information provided by the data is determined by its quality. It depends on the crystal itself, but also on the rigour of the data collection and analysis. This is a critical step because an error at this stage will affect the rest of the work and can slow down the determination of the structure. It is important to remember that the choice of space group made during the data processing is an assumption that may be invalidated later on during the phasing or refinement step. Similarly, if there is the slightest suspicion that there is an anomalous signal, it is advisable to assume that Friedel's law is not being respected and to use a file format that allows this signal to be exploited by separating the F^+ from the F^-.

Related videos

Data acquisition: the collection strategy (Acquisition des données : la stratégie de collecte)
https://youtu.be/EUHxyiAMzXo

Processing of diffraction data (Le traitement des données de diffraction)
https://youtu.be/AnVEJ_oSq1I

Diffraction data analysis (L'analyse des données de diffraction)
https://youtu.be/jqFEnTkGbkE

Chapter 8

Fourier Transform

8.1 Introduction to the Fourier Transform

A mathematical tool invented by Joseph Fourier in the early 19th century allows crystallographers to determine atomic structures from diffraction data. We have seen that a diffraction data set consists of peaks whose positions depend on the crystal lattice and whose intensities depend on the three-dimensional structure of the crystallised molecule. Furthermore, after analysing the diffraction data, the experimental result is an electron density map, which represents the number of electrons per unit volume at each point x, y, z of the crystal lattice. This map is used to build the three-dimensional model, including the spatial coordinates of each atom of the macromolecule under investigation. The diffraction data are in reciprocal space, with each Bragg spot located at the position $h\mathbf{a}^* + k\mathbf{b}^* + l\mathbf{c}^*$ (see §3.4). The electron density map is in direct space. To go from reciprocal space to direct space, or from direct space to reciprocal space, there is a mathematical operation called the Fourier transform, named after the mathematician who developed it (figure 8.1).

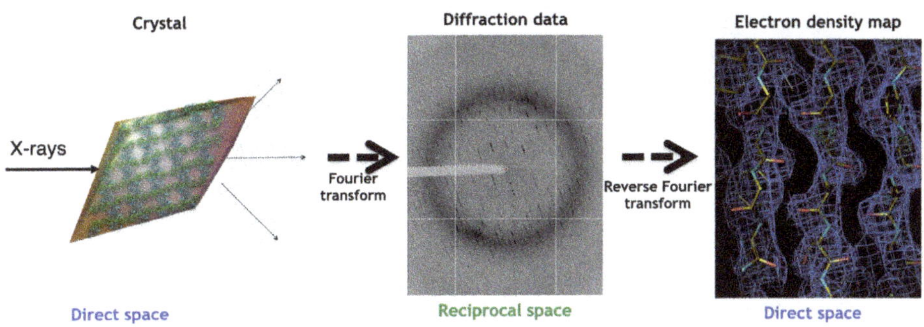

FIG. 8.1 – Fourier transform from reciprocal space to direct space.

DOI: 10.1051/978-2-7598-3501-0.c008

In the early 19th century, Joseph Fourier stated that any periodic function, no matter how complex (such as the electron density in a crystal), can be expressed as the sum of simple periodic functions such as sines or cosines (figure 8.2). What does this mean and why is it useful in crystallography?

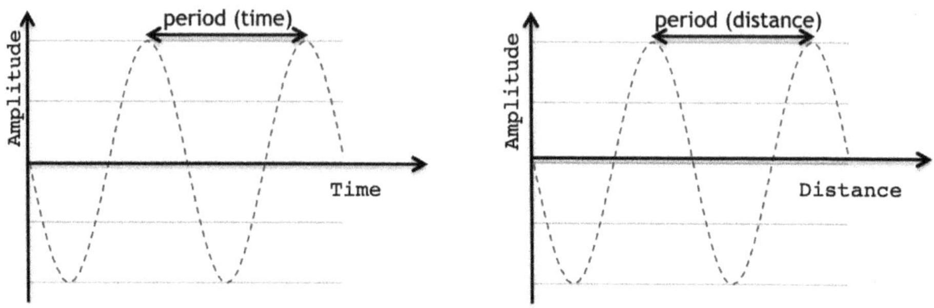

FIG. 8.2 – Periodic function in time (left) or space (right).

8.1.1 Fourier Transform Applied to Music

When we hear a musical note, such as an A, played by a flute or a violin, we realise that it is the same note. But the sensation is very different from one instrument to another. The flute has a pure sound, while the violin has a more complex sound. You can see this by recording the sound of the flute or the violin on an oscilloscope. The oscilloscope shows the wave associated with the sound. In both cases, there is a repeating pattern, but the pattern varies from instrument to instrument. In the case of the flute, it is a simple repetitive pattern consisting of a sine wave that represents the amplitude of the sound as a function of time. The frequency of 440 Hz means that this sine wave is repeated 440 times per second. The time span between two successive identical sinusoids is the period, which is the inverse of the frequency, in this case, 1/440 s (figure 8.3).

As it is a simple sine wave, it can be described by the time-dependent function:

$$F(t) = a \times \cos(v2\pi t) \tag{8.1}$$

a is the amplitude, to which we can arbitrarily assign the value of 1;
v is the frequency: 440 Hz in this case.

In the case of the violin, the pattern is more complex. The overall frequency is 440 Hz, which corresponds to the note A. The simple sinusoidal curve represents the so-called fundamental. We can also see that each repeated signal contains 6 peaks. To describe the violin signal, we therefore need to add at least a second sinusoid. We divide the fundamental into six so that we have six peaks instead of one (figure 8.3). This function is called the sixth harmonic and is written as:

$$f_6 = a_6 \times \cos(6v2\pi t) \tag{8.2}$$

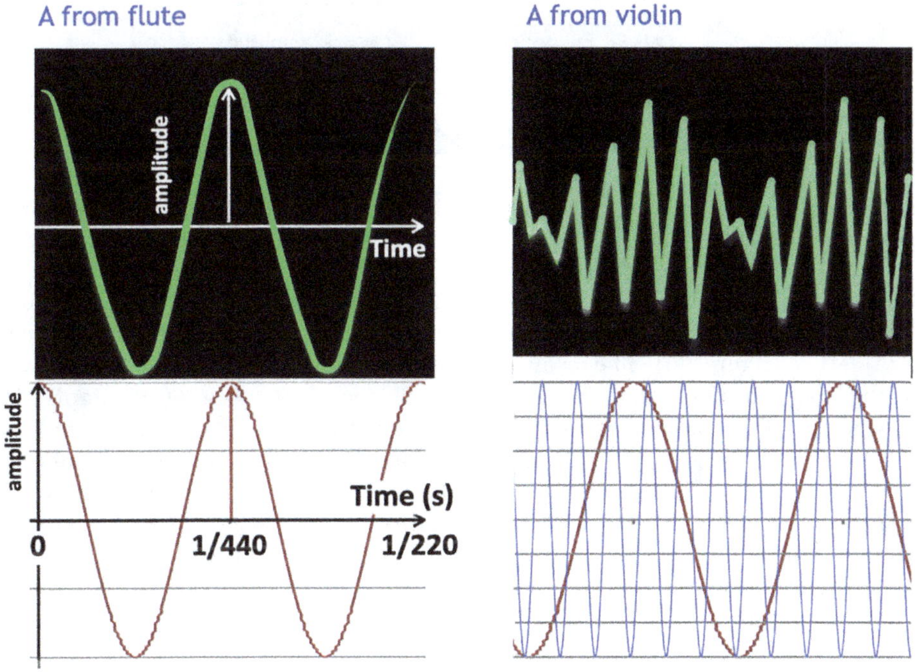

FIG. 8.3 – Representation of the sine curve associated with the note A played by a flute (left) or by a violin (right).

Is it possible to represent the periodic curve corresponding to the sound of the violin by adding a series of sine curves with periods of 1, 1/2, 1/3, 1/4, ...?

Joseph Fourier replied: Yes, it is possible.

8.1.2 Fourier Transform Applied to Crystal

It is possible to express a periodic function as a function of time, as shown above for music. In this case, the signal is expressed as a function of time. The period is the duration associated with the repetition of the signal (figure 8.2, left). A periodic function can also be expressed in space. The period in space is the minimum repetition interval (figure 8.2, right). The frequency, or spatial frequency, is the inverse of the period. We will apply the Fourier transform to a crystal system with periodic functions of this form.

The size of a crystal is about 100 μm and the size of a cell is about 100 Å or 10 nm. Therefore, there are about 10 000 repetitions of the cell and its contents along one side of the crystal. The content of each cell can be represented by its average electron density. Thus, in a given direction, the electron density in the crystal is represented by a periodic curve with a period equal to the size of the cell (figure 8.4).

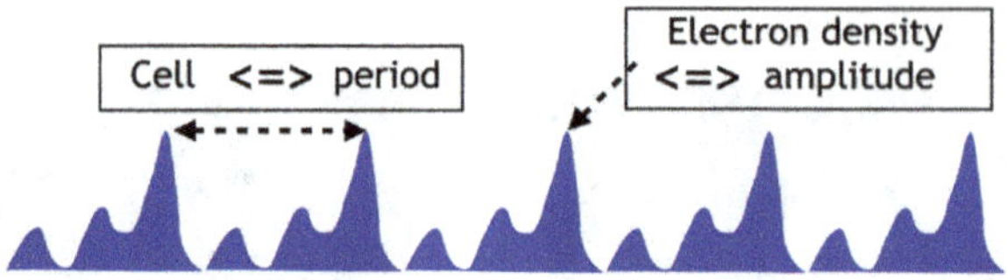

FIG. 8.4 – The equivalence between the repetition of the contents of the crystal cells and a periodic function.

Describing this periodic curve as a sum of sinusoids of different amplitudes is similar to applying a Fourier series development to the repetitions of an imaginary crystal lattice (figure 8.5). The mean value of the electron density is used to draw the straight line corresponding to f0, with an amplitude of a0 = 5 in our example. The first sine has the same period as the lattice, it is the f1 function. The curves corresponding to the successive harmonics are added one by one under the initial curve (on the left of the figure). The amplitude of each curve is represented by the diameter of the discs (right part of the figure). We can see that the sum of the successive harmonics allows us to describe the initial curve. If we wanted to reproduce this curve exactly, we would need an infinite number of sine waves. We can make a good approximation with a finite number. The higher the number of harmonics, the better the accuracy and the higher the resolution, *i.e.* the finer the details. The amplitudes shown in the right part of the figure are known: they are directly related to the intensity of the measured diffraction spots (figure 8.5).

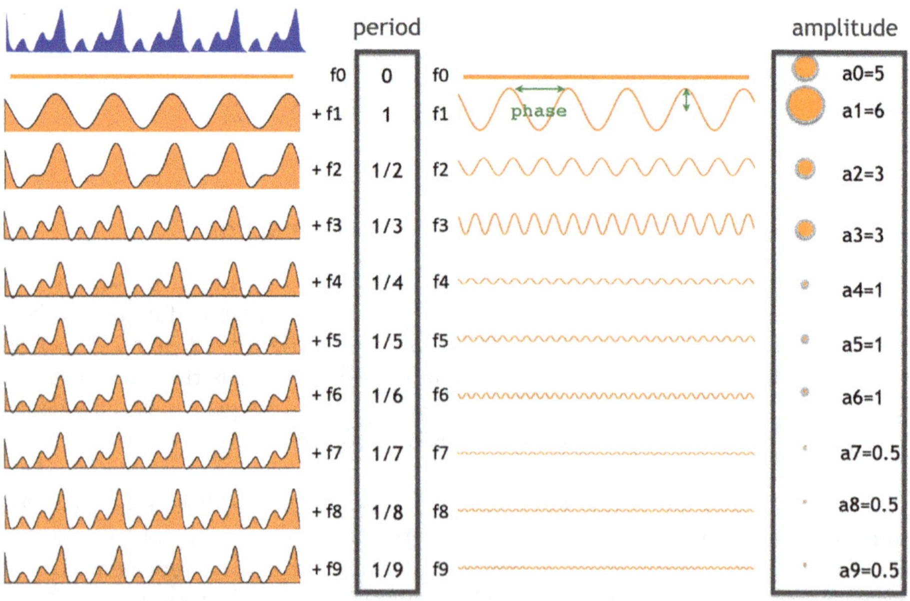

FIG. 8.5 – Fourier series development of the electron content of a crystal lattice. The diameter of the circles represents the amplitude of the sinusoids.

What would happen if we wanted to offset, *i.e.* phase shift, the first sine wave? Although we keep all the other sinusoids the same, the result is different from what we expect (figure 8.6).

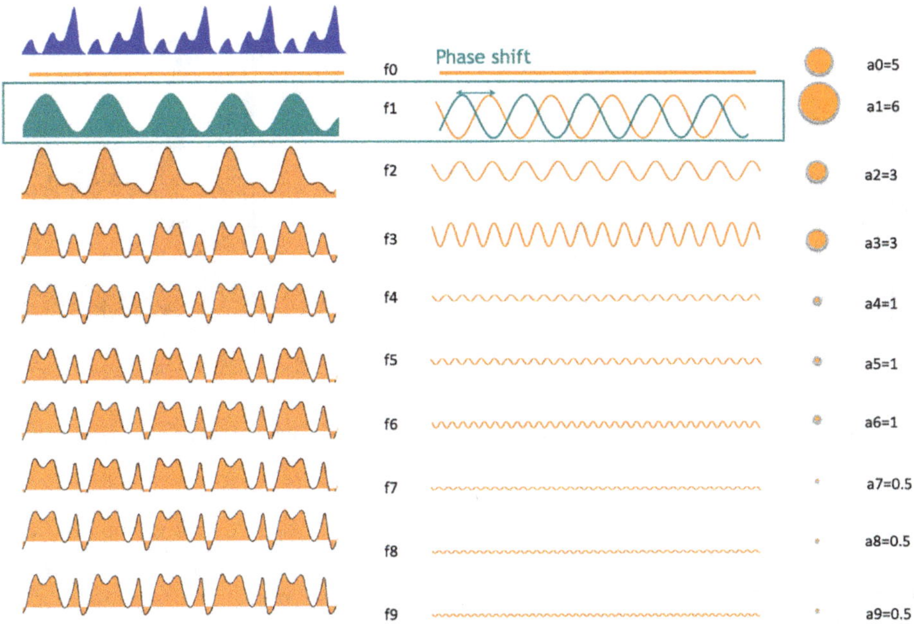

FIG. 8.6 – Effect of introducing a phase shift on f1 (green box) in the Fourier series development.

This means that the position of the curve, its phase, is an essential parameter in describing the complete signal. Therefore, if the period (or spatial frequency) associated with a wave is known, two pieces of information allow the signal to be reconstructed: the amplitude and the phase associated with each sinusoid (figure 8.7).

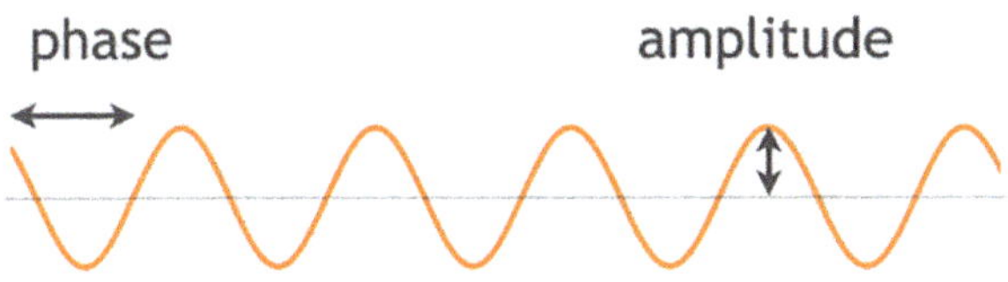

FIG. 8.7 – The two key elements for reconstructing a wave: phase and amplitude.

Each wave, described by an amplitude $|F|$ and a phase α, can also be represented in complex space. The amplitude is represented by a vector whose length is the modulus of F, *i.e.* the amplitude. The position of the vector depends on the phase alpha, which ranges from 0 to 2π (figure 8.8).

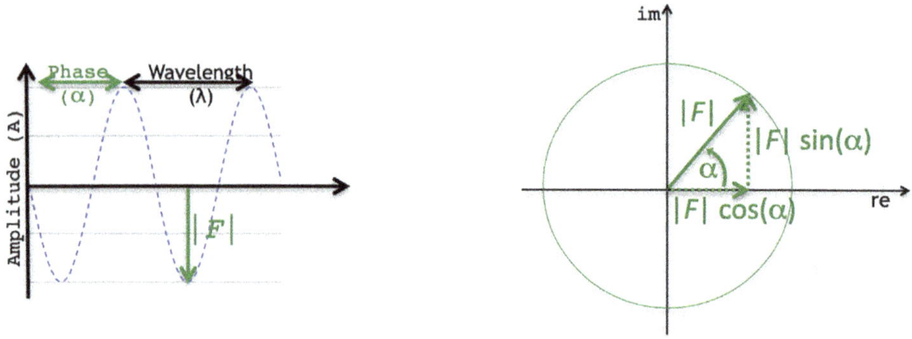

FIG. 8.8 – Representation of a sine wave (left) in complex space (right). re: real axis, im: imaginary axis.

This shows that a periodic curve can be described by a series of waves, represented by an amplitude, here proportional to the diameter of the point, and a phase, represented by their colour (figure 8.9). This series of waves is called the Fourier spectrum and corresponds to the Fourier series development of the periodic curve. The transition from the periodic curve to the Fourier spectrum, or from the spectrum to the curve, is a mathematical operation called the Fourier transform.

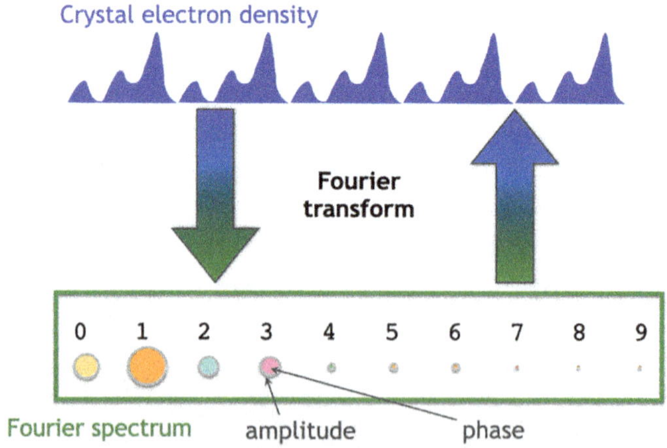

FIG. 8.9 – Schematic representation of the Fourier transform. In the Fourier spectrum, the diameter of each circle represents the amplitude of the peak. The colour symbolically represents the phase of the waves in the series.

In this example, we are working in one dimension only. As a crystal is three-dimensional, there is an electron density at each point x, y, z of the crystal lattice, and the Fourier spectrum becomes a set of diffraction spots.

8.2 The Fourier Transform and the Phase Problem

As we saw in chapter 1, Lawrence Bragg described the phenomenon of diffraction at the beginning of the 20th century in the form of the famous law named after him. This led to a minor disagreement with his father. Sir William Henry Bragg soon realised that his son was right and, more importantly, that the diffraction of X-rays by a crystal could be represented by a Fourier transform (figure 8.10). It is precisely because of this relationship between diffraction and the Fourier transform that the phenomenon of diffraction can be used to understand the spatial organisation of molecules.

FIG. 8.10 – The role of Bragg's father and son in using the phenomenon of diffraction to resolve the three-dimensional structure of a molecule.

8.2.1 The Fourier Transform

When a crystal diffracts an X-ray beam, its electron density ρ at each point x, y, z of the elementary cell can be described as the sum over the coordinates h, k, l of the moduli of the structure factors multiplied by the complex exponential of the phase associated with each factor, all related to the cell volume V:

$$\rho(xyz) = \frac{1}{V}\sum_h \sum_k \sum_l |F(hkl)|\, e^{i\alpha(hkl)}\, e^{-2\pi i(hx + ky + lz)} \tag{8.3}$$

V: cell volume;
$|F(hkl)|$: amplitude of the hkl reflection;
$\alpha(hkl)$: phase of the hkl reflection.

The indices h, k, l of the structure factors and the x, y, z coordinates of a point in the crystal are found in this mathematical expression of the Fourier transform.

Using the Fourier series development of the previous paragraph in a one-dimensional system, the electron density is then expressed as the sum over h:

$$\rho(x) = \frac{1}{L}\sum_h |F(h)|\, e^{i\alpha(h)}\, e^{-2\pi ihx} \tag{8.4}$$

In this example, the density $\rho(x)$ is the sum of the harmonics from f0 to f9 (figure 8.5), of which each wave or structure factor $F(h)$ can be described by its amplitude $|F(h)|$ and its phase $\alpha(h)$, related to the length of the cell L. In a three-dimensional system it is the same, $|F(hkl)|$ is the amplitude of the sine wave, $\alpha(hkl)$ its phase and $d(hkl)$ its period. Because of the three dimensions, the sine wave has a 'direction', that of the vector $h\mathbf{a}^* + k\mathbf{b}^* + l\mathbf{c}^*$. Each structure factor represents a wave described by an amplitude $|F(hkl)|$ and a phase α. To represent such a wave, it is better to use the complex plane (figure 8.8).

We can write:

$$F = |F|\cos(\alpha) + i|F|\sin(\alpha) \tag{8.5}$$

The Euler formula:

$$e^{i\alpha} = \cos(\alpha) + i\sin(\alpha)$$

Allows one to write this function in an exponential form:

$$F = |F|\, e^{i\alpha} \tag{8.6}$$

This notation gives access to the algebraic properties of the exponential function, which helps manipulate this expression. Considering the harmonic development of a wave, each harmonic f0 to f9 depends on the overall shape of the blue pattern (figure 8.5), which is given by:

$$F(h) = \int_V \rho(x)\, e^{2\pi i(hx)}\, dV \tag{8.7}$$

These two expressions, $\rho(x)$ (equation (8.4)) and $F(h)$ (equation (8.7)), highlight a key property of the Fourier transform, namely that each point of the density $\rho(x)$ is related to the sum of the harmonics of the Fourier series development, and each harmonic is related to the overall shape of the blue pattern, *i.e.* the total electronic content of the cell.

During the acquisition of an X-ray diffraction dataset from a crystal, the diffracted intensities $I(hkl)$ are recorded in three dimensions. Each diffracted intensity is the result of scattering from all the atoms in the cell. Each crystal cell has a volume V and at each point x, y, z of the cell the electron density depends on the atomic content. Each diffracted intensity is proportional to the square of the amplitude of the structure factor, written as $|F(hkl)|$. Therefore, each structure factor $F(hkl)$ results from the scattering of all the atoms contained in the cell (figure 8.11) and formula (8.7) can be written in three-dimensional form. The integral function is similar to a sum but takes into account the continuity of the electron density in the cell: in this form, it is the 3D Fourier transform of the electron density $\rho(xyz)$. This formula expresses the relationship between the structure factor in reciprocal space and the electron density in the crystal. It is important to understand that each atom contributes to the total structure factors (figure 8.11).

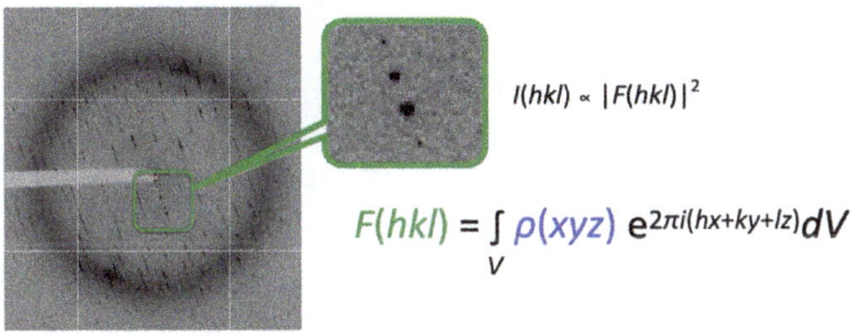

FIG. 8.11 – Fourier transform applied to the structure factors of a diffraction data set.

Conversely, each *hkl* index reflection in a diffraction data set contains part of the information about the electron density at each point x, y, z in the crystal lattice. This is the inverse Fourier transform:

$$\rho(xyz) = \frac{1}{V}\sum_h \sum_k \sum_l |F(hkl)|\, e^{i\alpha(hkl)}\, e^{-2\pi i(hx+ky+lz)} \tag{8.8}$$

V: cell volume;
$|F(hkl)|$: amplitude of the *hkl* reflection;
$\alpha(hkl)$: phase of the *hkl* reflection.

This is a sum – not an integral – because structure factors are discrete entities.

8.2.2 The Phase Problem

The recorded data correspond to the structure factors $F(hkl)$. The density $\rho(xyz)$ is unknown. The module of the structure factors $F(hkl)$ is the only known data. The phase is unknown. This is known as the phase problem. To solve the three-dimensional structure of a molecule we need to find the phases associated with the structure factors so that we can calculate an electron density map at each point x, y, z of the crystal lattice.

Several strategies have been developed to solve this phase problem, depending on the information we already have about the molecule of interest. For example, in the case of a protein, the three-dimensional structure of a similar protein can be available and therefore used as a model. If the crystal form (described by the lattice parameters, the crystal system and the position of the protein) of the protein under investigation is the same as the model, one could use the measured amplitudes and the phases from the model to apply a Fourier synthesis and calculate the electron density map. This can be done when studying mutants of a protein or complexes between a protein and small ligands. Otherwise, the crystal form is generally different from the model and the phasing method is the molecular replacement, which consists of positioning the model in the new crystal lattice. Finally, if a similar structure does not exist, a number of experimental phase methods can be used to resolve the phases (figure 8.12).

Identical or homologous protein	Crystal form	Phasing method
Known Structure	Identical	Fourier synthesis
	Different	Molecular Replacement
No neighbouring structure known		Single Isomorphous Replacement (SIR)
		Single wavelength Anomalous Dispersion (SAD)
		Multiple Isomorphous Replacement (MIR)
		Multi wavelength Anomalous Dispersion (MAD)

FIG. 8.12 – The various methods of 'phasing' a macromolecule structure.

Related videos

Introduction to the Fourier transform (Introduction à la transformée de Fourier)
https://youtu.be/mZ2fDZb9l2o

The Fourier transform and the phase problem (La transformée de Fourier et le problème de phase)
https://youtu.be/edSaxcHdTBU

Chapter 9

The Patterson Function

9.1 The Phase Problem and the Patterson Function

In 1935, Arthur Lindo Patterson visited the laboratory of the crystallographer Bertram Eugene Warren at the Massachusetts Institute of Technology. Patterson wondered what would happen if the Fourier transform were applied directly to the intensities, *i.e.*, the squared amplitudes, rather than to the structure factors:

$$I(hkl) = |F(hkl)|^2 \tag{9.1}$$

The Patterson function is therefore written as a Fourier series where the coefficients are the intensity $I(hkl)$ rather than the unmeasured structure factors $F(hkl)$. Patterson's idea was to use only the information available from the experimental data (figure 9.1).

$$\rho(xyz) = 1/V \sum_{hkl} |F(hkl)|\ e^{i\alpha} e^{-2\pi i(hx+ky+lz)}$$

Arthur Lindo Patterson

$$P(uvw) = 1/V \sum_{hkl} I(hkl)\ e^{-2\pi i(hu+kv+lw)}$$

FIG. 9.1 – Patterson function.

9.2 Properties of the Patterson Function

After some algebra, substituting the Fourier transform formula (figure 8.11) into the definition of the Patterson function above, we get:

DOI: 10.1051/978-2-7598-3501-0.c009

$$P(uvw) = \int_V \rho(xyz)\rho(u+x, v+y, w+z)dV \tag{9.2}$$

This is the autocorrelation function of the electron density.

9.2.1 First Property

The above formula shows that $P(uvw)$ will be maximum for values of u, v and w where $\rho(xyz)$ and $\rho(u+x, v+y, w+z)$ are maximum, *i.e.* at atomic positions.

So, if we consider two atoms 1 and 2 at the positions $x1, y1, z1$ and $x2, y2, z2$ in the crystal lattice, then $\rho(xyz)$ will have peaks at these points. At the coordinate vector u, v, w, the Patterson function has a peak such that:

$$u = x2 - x1$$

$$v = y2 - y1$$

$$w = z2 - z1$$

The norm of this vector is equal to the distance between atoms 1 and 2 (figure 9.2) (see box on the **Vector operations**, chapter 3).

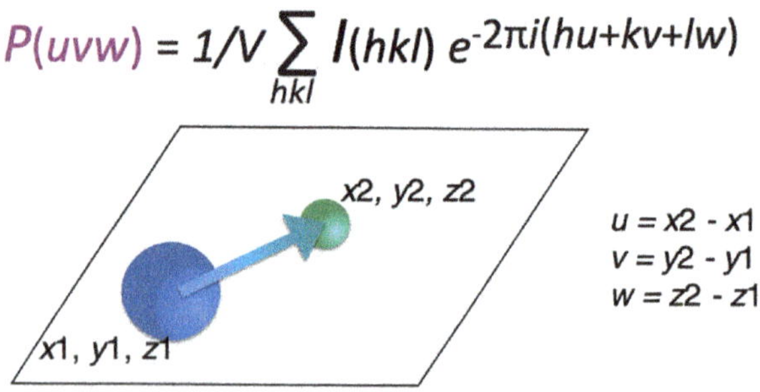

FIG. 9.2 – The Patterson function and its representation. The blue arrow represents the u, v, w coordinate vector in Patterson space, joining atoms 1 and 2.

9.2.2 Second Property

Since the Patterson function is periodic with period 1 in all three directions ($P(u+1, v+1, w+1) = P(uvw)$), the crystal space and the Patterson space have the same cell.

We will use this property to build a step-by-step example of a Patterson function. Consider a crystal lattice containing three atoms with coordinates $x1, y1, z1$; $x2, y2, z2$; and $x3, y3, z3$ (figure 9.3a). The interatomic vectors of these three atoms form a triangle (figure 9.3b). To represent the Patterson function, we use the crystal cell with the three atoms 1, 2, and 3 (figure 9.3c). We place each corner of the triangle successively at the cell origin (figure 9.3d–f). As the crystal is periodic, we repeat this operation for each cell (figure 9.3g–h). For a given cell we now have a set of peaks, each representing an interatomic distance (figure 9.3i). Thus, we find the peaks corresponding to vectors whose norms are equal to the interatomic distances (figure 9.3j). For example, $u = x2 - x1$, $v = y2 - y1$, $w = z2 - z1$, are the coordinates of the vector connecting atom 2 to atom 1.

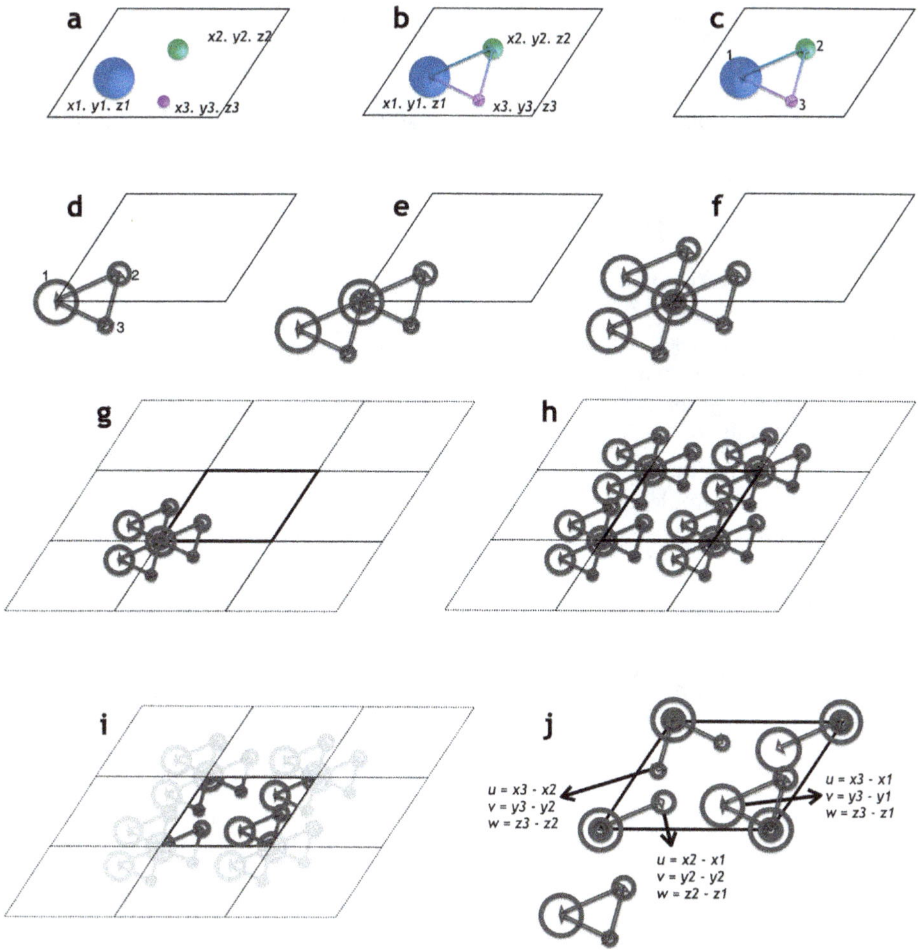

FIG. 9.3 – Cartoon showing the Patterson function of a cell containing three atoms.

For a cell containing N atoms, the number of peaks to be found in the Patterson map will be N multiplied by N to include all atom-to-atom distances, minus N because the distance of an atom to itself is zero, corresponding to the peak at the origin. This gives a peak number of $N^2 - N$ (figure 9.4). With three atoms: $3^2 - 3 = 6$ peaks.

As the number of atoms increases, the number of peaks quickly becomes very large.

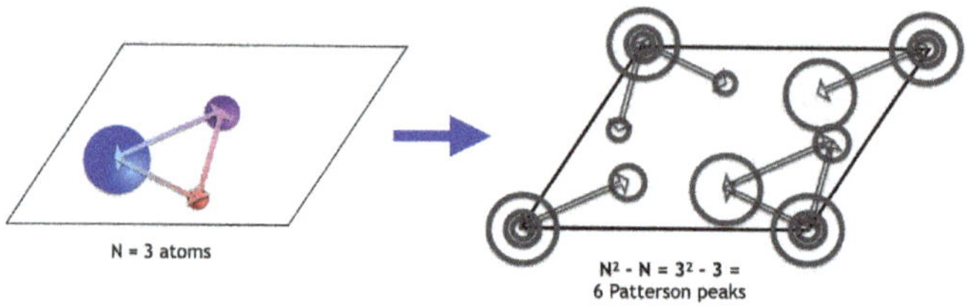

FIG. 9.4 – Calculation of the number of interatomic peaks in a Patterson map.

9.2.3 *Third Property*

An important property of the Patterson function is that the height of the peaks is proportional to the product of the number of electrons in the atoms.

This property is used to detect the presence of heavier atoms in a structure. Because an atom is not a point, it has a certain volume which can lead to overlapping of Patterson peaks. For example, if two peaks are very close together, a plot of their heights will show an area of overlap. In this area, the two peaks merge into a single broader peak (figure 9.5).

In biology, macromolecules are made up of thousands of atoms. As a result, the Patterson map has a very large number of peaks, resulting in a lot of overlaps.

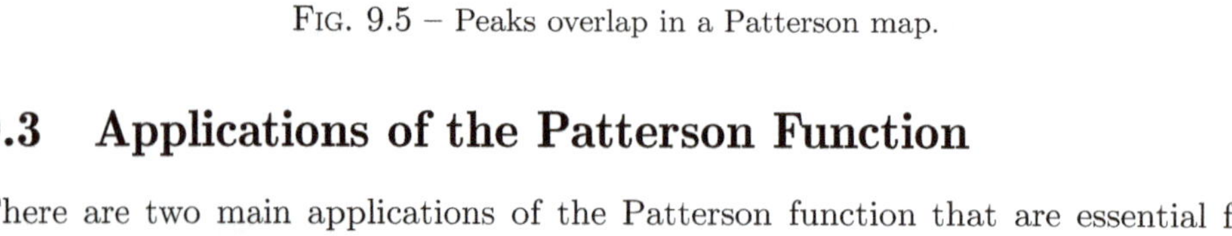

FIG. 9.5 – Peaks overlap in a Patterson map.

9.3 Applications of the Patterson Function

There are two main applications of the Patterson function that are essential for solving the phase problem:

- The molecular replacement method consists of positioning a model of a known structure in a new crystal lattice (see chapter 10). The Patterson function allows the model orientation to be separated from the model positioning step, which greatly speeds up the calculation time.

- The experimental phasing approaches, whether isomorphic replacement or anomalous scattering signal analysis, all start by determining the position of the heavy atoms or the anomalous scatterers. This is done by calculating the differences between the data sets so that only the vectors of interest are retained (see chapter 11). Patterson maps calculated from these intensity differences allow the position of the heavy atoms or anomalous scatterers to be resolved and provide a first estimate of the phases.

Related videos

The Patterson function (La fonction de Patterson)
https://youtu.be/gyxrLUHtGNo

Chapter 10

Molecular Replacement

The Fourier transform and the Patterson function are the main theoretical tools for solving the three-dimensional structure of a macromolecule. We have seen that introducing a phase shift on a single harmonic of the series no longer allows us to reconstruct the original signal (figure 8.6). The relative importance of the phases can be illustrated by calculating an 'electron density map' using the amplitudes from a duck diffraction and the phases from a cat diffraction. The resulting image of a cat shows the dominant weight of the phases (figure 10.1).

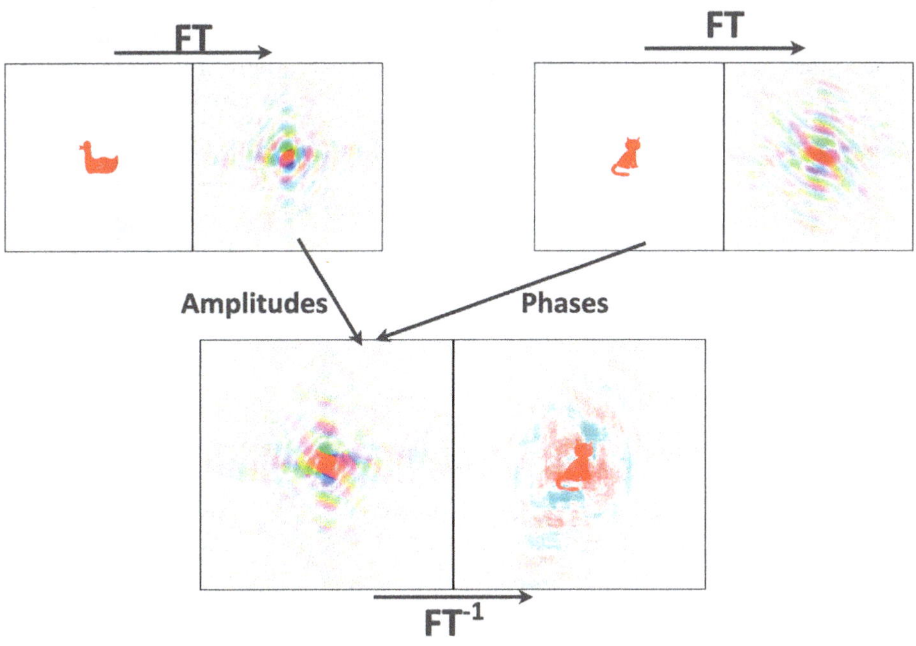

FIG. 10.1 – The cat and duck phasing, borrowed from Kevin Cowtan (Book of Fourier, http://www.ysbl.york.ac.uk/~cowtan/fourier/fourier.html).

DOI: 10.1051/978-2-7598-3501-0.c010

Since phases are dominant and cannot be measured, mathematical formalisms have been developed to estimate them. The two main approaches are molecular replacement and experimental phasing.

10.1 Molecular Replacement

In some cases, it can be assumed *a priori* that the structure of the molecule under investigation is close to that of a molecule with a known structure. For example, the bibliographic research associated with the project may indicate the three-dimensional structure of a protein that is related to the protein, part of the protein or the functional assembly under investigation. Another possibility is to identify the structure of a protein with high sequence similarity to the protein under investigation. In both cases, the known structure is available in the Protein Data Bank or PDB (see Introduction). The known structure can be used as a 'template' to solve the phase problem by molecular replacement.

10.1.1 Patterson Function and Molecular Replacement

The purpose of molecular replacement (MR) is to position a three-dimensional model in the crystal system under investigation. An experimental Patterson function is calculated from the experimental data. The peaks of this Patterson function correspond to the interatomic vectors (figure 10.2). Some of the interatomic vectors are intramolecular and represent the distances between atoms of the same molecule

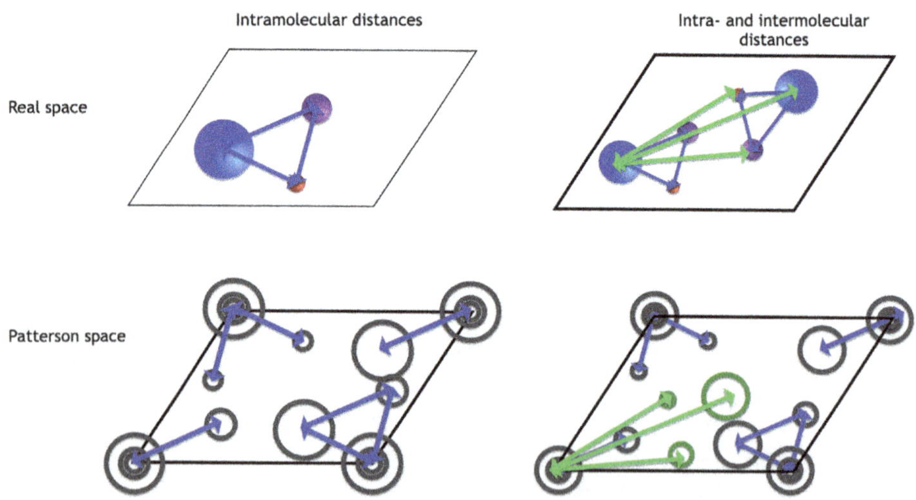

FIG. 10.2 – Representation of the vectors associated with intramolecular (blue) and intermolecular (green) distances, in real or Patterson space. For clarity, only the intermolecular vectors within a cell are shown.

in the crystal. Their position depends on the orientation of the molecule in the lattice. Other interatomic vectors are intermolecular, *i.e.* they represent the interatomic distances between different molecules in the crystal. Their position depends on the arrangement of the molecules relative to each other in the lattice (figure 10.2).

It is possible to dissociate the information about the orientation of the molecule in the cell (provided by the intramolecular vectors) from the information about the position of the molecule in the cell (provided by the intermolecular vectors) because the intramolecular interatomic distances are shorter than the intermolecular interatomic distances.

The first step is to describe the orientation of the model by a rotation operation. To do this, we first place the model in an artificial cell with no symmetry operation and whose parameters are large enough to prevent the presence of any neighbouring molecules. Using this theoretical crystal, we apply the Fourier transform to calculate the theoretical diffraction intensities. We then use these theoretical intensities to calculate the Patterson function. Due to the large size of the box and the presence of a single molecule in the cell, the peaks in the theoretical Patterson map correspond to a projection of intramolecular vectors. Their position depends on the orientation of the molecule in this artificial cell. To find out the orientation of the molecule in the cell, we superimpose the Patterson map calculated with the theoretical data on the Patterson map calculated with the experimental data. We rotate the Patterson map sequentially and at each angular position, we calculate the correlation between the two Patterson maps. This is the rotation function. When the maxima corresponding to the intramolecular vectors of the two Patterson functions overlap, the correlation is maximal. This means that the Patterson function calculated for that particular rotation describes the orientation of the molecule in the experimental crystal cell. The orientation can then be deduced directly (figure 10.3).

Once the orientation of the molecule has been obtained, the second step is to determine its position in the experimental crystal lattice, which corresponds to the translation of the molecule with respect to the origin. For each possible position of the molecule in the lattice, the Patterson function is calculated and compared with the experimental data. This is done using the symmetry operations of the crystal. The intermolecular vectors depend on the relative positions of the molecules, which depend on the symmetry operations (figure 10.4). Since we know the orientation of the model, we can calculate the position of the molecules that are symmetrically related to it. The translation function mathematically expresses the correlation between the calculated and experimental Patterson.

At this stage, the maximum correlation peak reflects both the correct orientation and position of the molecule and represents the molecular replacement solution. Without prior knowledge of the orientation of the molecule, the search for the position would be laborious. It would be necessary to search for the correct orientation for each possible position of the cell. Therefore, the strength of this molecular replacement method lies in the decoupling of the rotation and the translation steps, which allows the six parameters to be searched three by three.

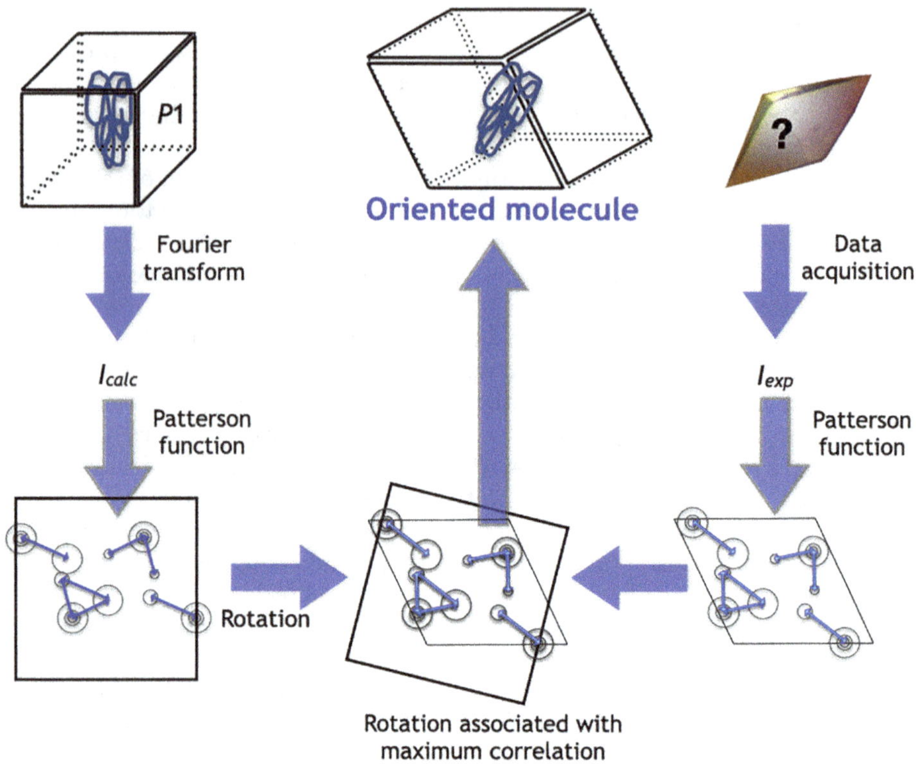

FIG. 10.3 – Use of Patterson function to orient the molecule in the cell.

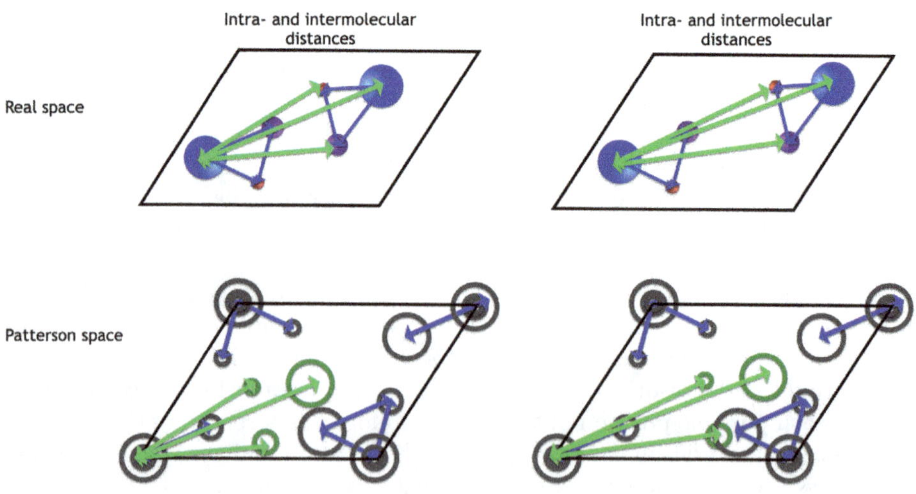

FIG. 10.4 – Effect of the position of the molecules on the intermolecular interatomic vectors.

10.2 Evaluation of the Molecular Replacement Result

No matter what the molecular replacement problem is, no matter what software is used, the result will always be in the form of a list of correlation coefficients. In order to identify the correct solution, which is not necessarily the one with the highest correlation coefficient, it is essential to take into account several criteria.

The first factor is the quality of the diffraction data:

- The accuracy and reliability of the measurements are an essential parameter already discussed in chapter 7. Poor data quality is indicated by a high R-factor or a low signal-to-noise ratio.
- The completeness of the data, which corresponds to the percentage of data collected relative to the number of possible reflections at a given resolution. If the cell parameters and symmetry are known, it is possible to predict how many reflections can be recorded. If part of the reciprocal space is missing (*i.e.* uncollected angular area), the effect is greater than if the missing data is evenly distributed.

The second factor comes from the model quality:

- Its homology with the molecule under study: generally, below 30% sequence identity, molecular structures can be very different and the errors associated with the calculated intensities become too large for rotation or translation functions to find the correct solution.
- The fraction of the structure of interest that is represented by the model. The model may be very close to, or even identical to, the molecule being studied, but may only cover a small fraction of it. In this case, even if the molecular replacement solution is correct, the phases derived from it are not sufficiently accurate to calculate an electron density map to reconstruct the whole molecule. In this case, the phasing power provided by the model is considered insufficient.
- Finally, the quality of a model is determined by its structural homology to the target protein, not just by sequence identity. Only conformers of a given protein model with high structural homology to a target protein will provide a solution for molecular replacement.
- Similarly, the accuracy and reliability of the model affect the result. For example, the nature of the errors associated with older NMR structures makes them more difficult to use in molecular replacement, as do very low-resolution structures, regardless of the method used to determine them.

The third factor is the method used to evaluate the likelihood of the solutions found by molecular replacement. Classical methods use correlation coefficients such as the rotation and translation functions discussed above. More recently, statistical approaches have been developed to better deal with errors associated with the model or the data.

Finally, the fourth factor is the resolution of the experimental data, which has to be adapted to the model. In particular, the farther the model is from the structure under investigation, the less relevant the information provided by the high-resolution data will be in the search for a solution. On the other hand, if the space group of the crystal is of high symmetry, high-resolution data will allow the different solutions to be distinguished.

Related videos

Molecular replacement (Le remplacement moléculaire)
https://youtu.be/5w6EY-ofCIA

Chapter 11

Experimental Phasing

11.1 Isomorphous Replacement

If no molecular replacement model is available, the phase calculation requires the initial determination of a small part of the content of the molecule: one or more heavy atoms in the case of isomorphic replacement, or one or more anomalous scatterers in the case of the anomalous diffusion method.

11.1.1 Natives and Derivatives Data

Two properties allow isomorphous replacement:

1- Each recorded intensity provides a structure factor, which is related to all the electrons in the crystal cell by Fourier transform. Consequently, when the crystal content is modified, all the structure factors are affected.

2- Macromolecular crystals are porous. They contain an average of 50% solvent in the form of crystal channels. All kinds of small molecules can diffuse through these channels, and can potentially interact with the macromolecules that make up the crystal.

To take advantage of these properties, we collect a first dataset with a so-called 'native' macromolecular crystal in the absence of any modification. This data set corresponds to the native data. We then soak a crystal of the same form in a solution containing a soluble heavy atom salt with high electron density, such as platinum or mercury, which have 78 and 80 electrons respectively. In practice, we either immerse this crystal in the heavy atom salt solution so that it diffuses into the crystal, or we add the heavy atom salt to the crystallisation solution to co-crystallise the protein with the heavy atom. The heavy atom salt may interact with one or more chemical groups in the crystallised molecule. We collect a second dataset with this crystal, which we refer to as the derivative data (figure 11.1).

DOI: 10.1051/978-2-7598-3501-0.c011

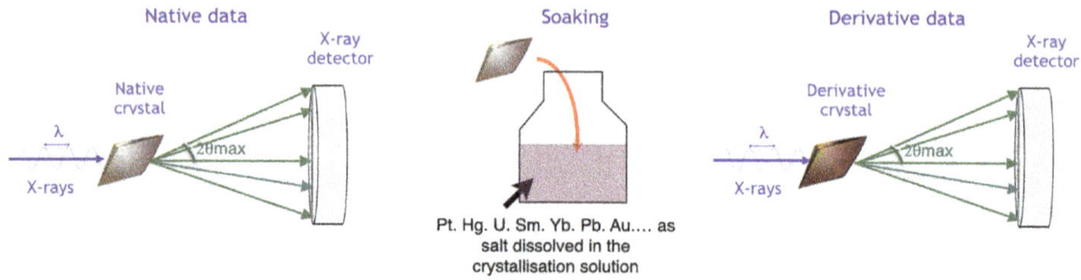

FIG. 11.1 – Acquisition of native data (left); preparation of a derivative crystal by soaking in the crystallisation solution supplemented with heavy atom salt(s) (centre), and acquisition of derivative data (right).

If the heavy atom has interacted in an ordered manner with certain macromolecular groups in the crystal, this will increase the number of electrons at certain positions. As a result, the intensities collected with the derivative crystal will be different from those of the native crystal. We can describe this phenomenon using a vector representation as in figure 11.2, left.

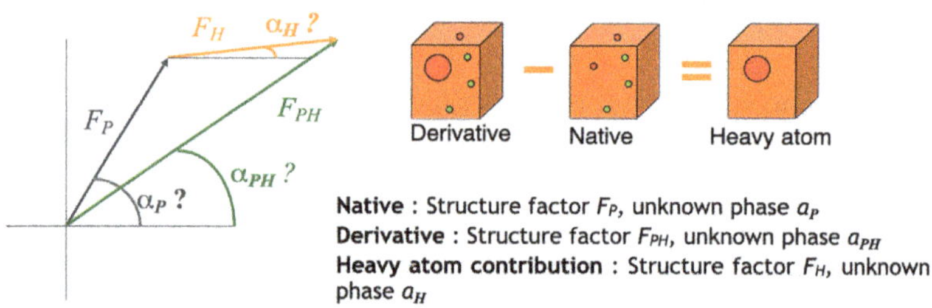

FIG. 11.2 – Vector representation of the native and derivative data (left) and illustration of the signal decomposition to locate the heavy atom (right).

To go from the native data to the derivative data, we plot the F_H vector, which also has an unknown phase at this stage. If we find the position H of the heavy atoms, we will know the electron density $\rho(xyz)$ of the heavy atoms H, and the F_H and α_{H} associated with each reflection of the reciprocal space can be calculated.

The Patterson function can be used to locate heavy atoms that occupy regular positions in each crystal cell. Two conditions are necessary for using the Patterson function:

1- The derived crystal must be isomorphic to the native crystal: the lattice parameters and the symmetry of the crystal must be preserved between the native and the derived crystal, and the molecule must not have undergone any rotation, translation or significant rearrangement of the lattice: the binding of the heavy atoms must not disturb the lattice content.

2- The contribution of these atoms to the diffracted intensities must be sufficiently strong and extended towards high resolution. To achieve this, they must bind exactly the same molecular site in each unit cell.

Both of these conditions can be assessed at the time of scaling the native and derived data[1], prior to calculating the isomorphic differences. At this stage, if the differences are very large over the whole resolution range, this may indicate a lack of isomorphism. Conversely, if the differences are too small, the data may not be suitable for site location and phase calculation.

11.1.2 Patterson Function and Harker Sections

Subtracting the native structure factors from the derived structure factors should give heavy atom structure factors:

$$|F_{PH}|^2 - |F_P|^2 = |\Delta Fiso|^2 \tag{11.1}$$

(iso for isomorphous).

This difference is an approximation of $|F_H|^2$ (figure 11.2).

By calculating a Patterson function with these differences, the symmetry elements of the crystal system help us to find the position of the heavy atom. For simplicity, we will consider the case where there is only one heavy atom per unit cell. When calculating the u, v, w positions of the Patterson space, the symmetry elements are associated with specific sections. These are called Harker sections. For example, in the case of the monoclinic space group $P2_1$, there are 2 equivalent positions: x, y, z and $-x, y+\frac{1}{2}, -z$. In Patterson space, to determine the coordinates of the interatomic vectors between symmetrically related atoms, we must calculate the u, v, w coordinates corresponding to symmetrical positions. The difference between the position x, y, z and itself is 0 0 0, as is the difference between the position $-x, y+\frac{1}{2}, -z$ and itself. The cross-results between x, y, z and $-x, y+\frac{1}{2}, -z$ provide the information. In the particular case of $P2_1$, the position $v = ½$ corresponds to a plane in Patterson space whose position is independent of the position of the heavy atom. This is a Harker section which depends only on crystal symmetry (the planes $v = ½$ or $v = -½$ are equivalent because we use fractional coordinates between 0 and 1 and because the Patterson function is centrosymmetric, as explained in chapter 9). In this section, we find two peaks due to the presence of the heavy atom. The u, w coordinates of the peaks give us directly the x and z of the corresponding atom. In this map, the two peaks observed mean that a heavy atom is bound in each cell of the crystal with which we have collected the derivative data set. At this stage, we do not know the y position, but we can set it to 0. Since we have located an atom, we can calculate its structure factor F_H, and we know the phase associated with F_H (figure 11.3).

[1]Scaling between two data sets consists of calculating proportionality factors between the data sets to take into account the experimental parameters (crystal size, scattering power, exposure time...).

Harker sections of space group $P2_1$:

u v w	x y z	-x y+1/2 -z
x y z	0 0 0	-2x +1/2 -2z
-x y+1/2 -z	2x -1/2 2z <=> 2x 1/2 2z	0 0 0

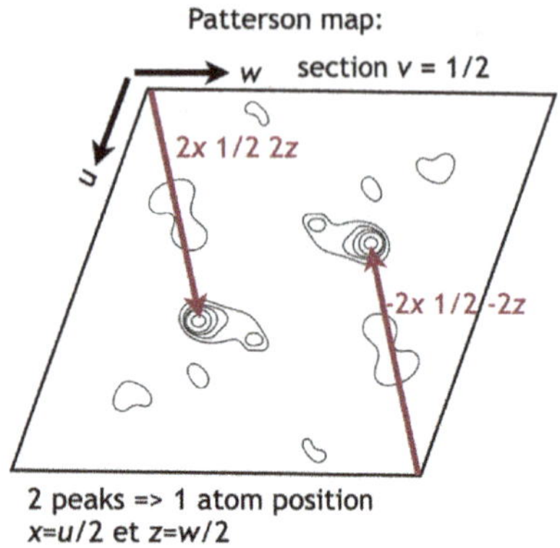

FIG. 11.3 – Calculation of Harker sections for space group $P2_1$ (left), and Patterson difference map between native and derivative data in space group $P2_1$ (right).

11.1.3 Vector Representation of Harker Section

With a single isomorphic derivative, we perform a single isomorphic replacement or SIR. For a given diffraction spot, we know the amplitude of the structure factor F_P, represented by the length of the vector, but we do not know its phase. To retrieve this information, there is a geometric operation called the Harker construction. We draw a circle representing all the possible values of the phase. From the origin, we can position the vector $-F_H$, which corresponds to a subtraction of the vector F_H. Knowing the position of the heavy atom, we know the associated phase, so we know the direction of the vector. At the end of the vector $-F_H$, we draw the vector F_{PH}, which corresponds to the size of the structure factor of the derived data. To represent the associated phase, which is unknown, we also draw a circle. We observe two points of intersection and therefore two possible positions of the vectors F_P and F_{PH}. Since we know that $F_{PH} - F_H = F_P$ (see the box on the **Vector operations** in chapter 3), the two intersection points correspond to the two possible positions for these vectors, and to the two possible phase values (figure 11.4, left).

So the information provided by a single heavy atom derivative leads to two possible phases for each structural factor. We have phase ambiguity. Let us add a second heavy atom derivative. We then use the Multiple Isomorphic Replacement (MIR) method. We have a second structure factor F_{H2} and a second structure factor F_{PH2}. A new circle is associated with the possible phases of F_{PH2} and intersects the F_P circle at two points corresponding to the two possible phases, as we have just seen. However, there is now only one point of intersection common to all three circles. This corresponds to the correct phase of the structure factor F_P. Therefore, by using a second heavy atom derivative, at a different position in the crystal lattice, the information provided allows us to overcome the phase ambiguity. The phase problem is solved and we can calculate an electron density map (figure 11.4, right).

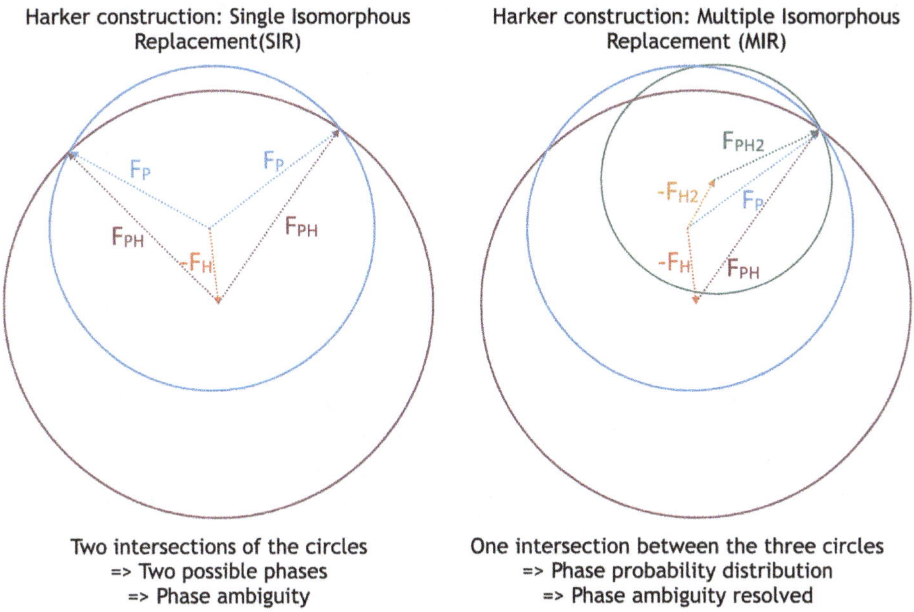

FIG. 11.4 – Vector representation of Harker sections and of the phase ambiguity in the SIR method (left), solved in the MIR method (right).

11.2 Anomalous Diffusion

A major advantage of producing X-rays with a synchrotron is that the wavelength of the radiation is tuneable. This means that the energy, or wavelength, of the X-rays can be tuned.

11.2.1 The Anomalous Signal

Under 'normal' conditions, X-ray scattering depends only on the electron density of the atom (§4.3). The scattering factor f is a real number, f_0, and at specific wavelengths the anomalous scattering phenomenon induces a significant variation in the scattering signal (figure 11.5).

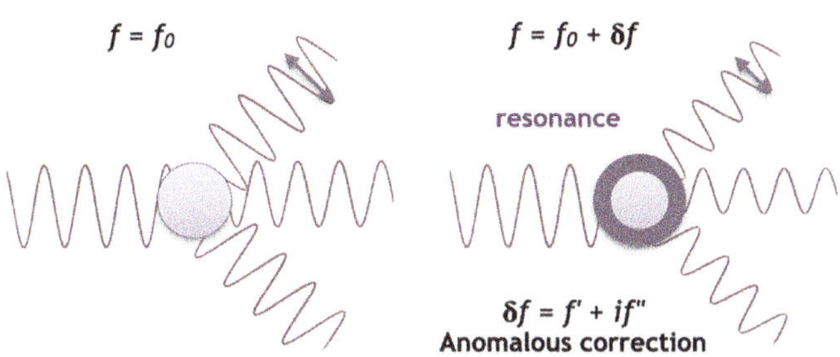

FIG. 11.5 – Scattering of an incident wave by an atom according to the number of electrons, and according to the wavelength.

We can write the scattering factor of the atom as:

$$f = f_0 + \delta f$$

δf is the anomalous correction. This is a complex number written as:
$\delta f = f' + if''$

The complete description of the scattering factor, f, includes three terms:

- f_0, which only takes into account the number of electrons;
- f', called the dispersive term;
- f'', which is related to the absorption phenomenon and is associated with the imaginary part.

f' and f'' depend on the wavelength, unlike f_0.

For a given element, the fluorescence emission signal as a function of X-ray energy shows a sharp increase at a particular energy (figures 7.3 and 11.6). As a result, the amplitude of the diffracted waves will be different at the energy corresponding to the maximum fluorescence – at the top of a peak where the f'' component is maximum, or at a distant energy where the scattering factor depends almost solely on the number

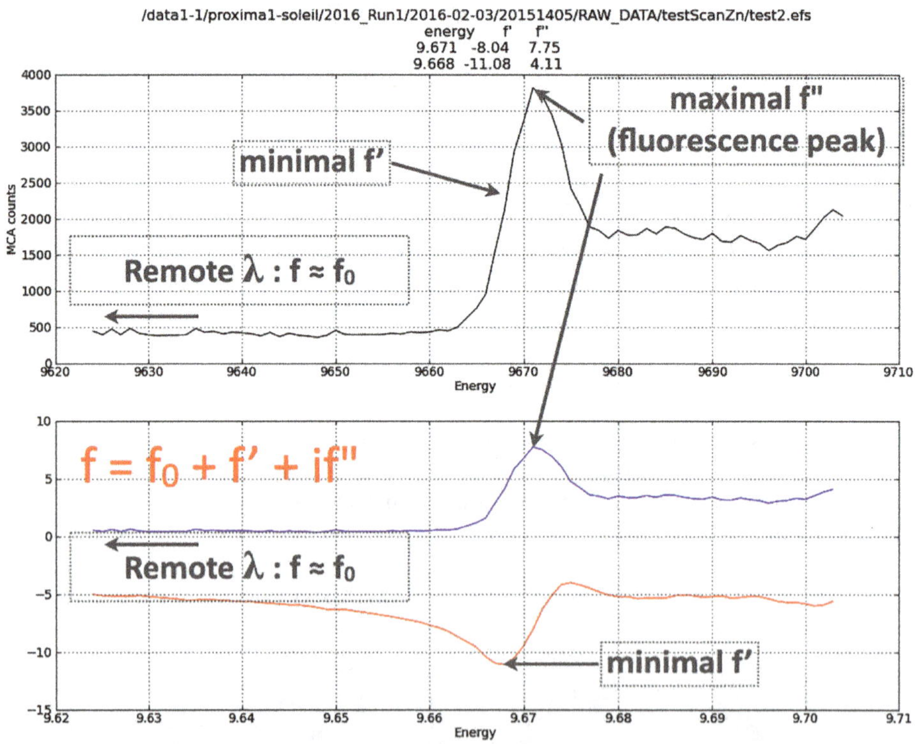

FIG. 11.6 – Variation of the intensity of the Zn fluorescence emission signal as a function of energy (top). The Kramer–Koenig equation links the experimental value of f' to that of f'' obtained by measuring the fluorescence of the crystal (bottom).

of electrons. This occurs at an energy that is characteristic of each atom. All atoms have at least one absorption edge and may have several. However, the energy or wavelength at which this phenomenon occurs is not always available on a beam line (for example, for S, P or Cl we would need energy below 3 keV), so we cannot always work at the absorption edge to solve the phase problem.

11.2.2 Breaking Friedel's Law

The structure factor $F(hkl)$ can be expressed:

$$F(hkl) = \sum_j f_j e^{2\pi i(hxj + kyj + lzj)} \tag{11.2}$$

where the sum is on the atoms j of the cell, at the position x_j, y_j, z_j. In the absence of anomalous scattering, *i.e.* when the scattering factors f_j are real, or when the crystal is centrosymmetric, equation (11.2) shows that:

$$F^*(hkl) = F(\overline{hkl})$$

Thus, we can write:

$$I(hkl) = F(hkl) \times F(\overline{hkl})$$

or:

$$I(hkl) = F(hkl) \times F^*(hkl)$$

In the same way, for the $\overline{hkl}$ indices:

$$I(\overline{hkl}) = F(\overline{hkl}) \times F(hkl)$$

or:

$$I(\overline{hkl}) = F^*(hkl) \times F(hkl)$$

As a consequence:

$$I(hkl) = I(\overline{hkl}) \tag{11.3}$$

This is Friedel's law.

Suppose that a crystallised macromolecule contains an anomalous scatterer H. For the non-derivative crystal:

$|F_P(hkl)| = |F_P(\overline{hkl})|$;

and the corresponding phases are opposite.

Similarly, in the absence of anomalous diffusion:

$|F_{PH}(hkl)| = |F_{PH}(\overline{hkl})|$;

and the corresponding phases are also opposite.

Between F_P and F_{PH}, we can draw F_H, which is also equivalent for *hkl* and for $\overline{hkl}$. Friedel's law is respected (figure 11.7, left). At a wavelength where the anomalous signal affects the scattering, there is an additional imaginary contribution, f'', corresponding to the anomalous signal of the H atom. This f'' contribution is always oriented at 90° from F_H. Consequently, the amplitude and phase of the vectors $F_{PH}(hkl)$ and $F_{PH}(\overline{hkl})$ are different. Friedel's law is broken (figure 11.7, right).

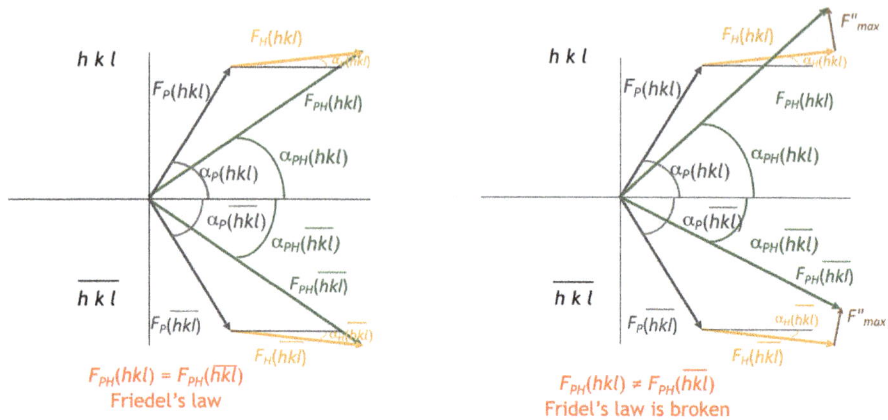

FIG. 11.7 – Vector representation of structure factors showing the respect (left) and violation (right) of Friedel's law.

11.2.3 The Use of Anomalous Signal

The location of an anomalous scatterer in the crystal cell allows us to solve the phase problem. As in the case of isomorphous replacement, we use the Patterson function to find the position of the anomalous scatterer. If the wavelength of the data is associated with a violation of Friedel's law, the Patterson function is applied to the difference between the structure factors measured at two different wavelengths, which is the difference between the *hkl* and $\overline{hkl}$ reflection intensities (figure 11.8). The advantage of using the anomalous signal is that the phase problem can be solved using data from a single crystal. Isomorphism is guaranteed because the signal is intrinsic to the crystal. The condition is the presence of a scatterer whose absorption edge is in the X-ray range, which is not the case for the so-called light atoms such as carbon, nitrogen or oxygen.

Sulfur atoms in methionine or cysteine residues have an anomalous scattering signal at a wavelength of about 5 Å, which is almost inaccessible on a beamline and presents many experimental problems. At this wavelength, X-rays are strongly absorbed by the air, the crystal and its support, making it difficult to measure the signal and separate it from the noise caused by absorption and scattering. In the case of nucleic acids, phosphorus can be used in a similar way to sulfur, with an absorption edge of about 5.8 Å. The ability to use these atoms has been greatly improved by the advent of pixelated detectors, whose low background makes it possible to extract weak signals.

The most common method is to replace some of the sulfur atoms in proteins with selenium atoms. It is possible to add selenomethionine instead of methionine to the culture medium when proteins are produced in bacteria. In this case, when the bacteria produce the protein, they incorporate selenomethionine instead of methionine (see chapter 2). The absorption edge of selenium is about 0.98 Å, with a strong anomalous signal, which is particularly useful for X-ray diffraction

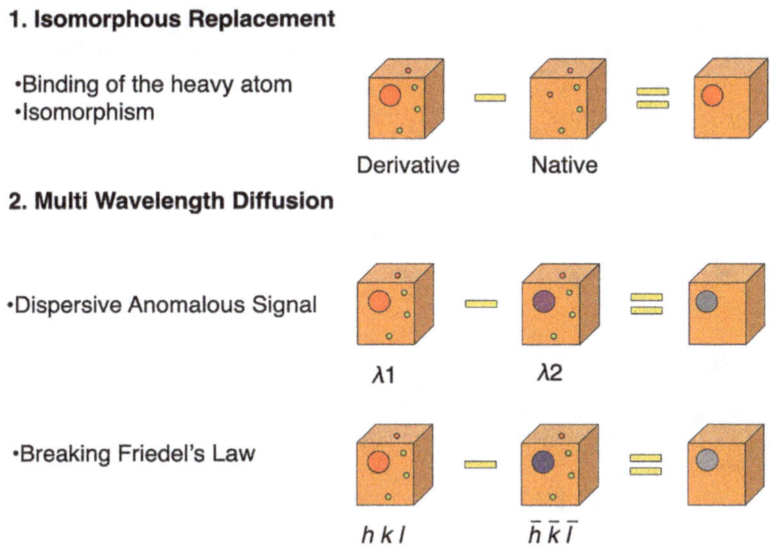

FIG. 11.8 – The use of the anomalous signal compared to that of isomorphous signal.

(see absorption edges of elements in the periodic table of the elements: http://skuld.bmsc.washington.edu/scatter/AS_periodic.html)

The formalisms described for isomorphic replacement are identical in the case of anomalous scattering. We refer to SAD (for Single Anomalous Dispersion) when we have data at a single wavelength. In this case, we collect the data at the wavelength corresponding to the fluorescence peak to take advantage of the breaking of Friedel's law. The phase ambiguity is similar to that of the SIR method. When we collect diffraction data at multiple wavelengths, we call it MAD (for Multiple Anomalous Dispersion) and there is no phase ambiguity. It is also possible to combine isomorphous replacement and anomalous dispersion. This is called SIRAS (Single Isomorphic Replacement with Anomalous Scattering) when using native data and a heavy atom derivative data collected at the fluorescence peak of the heavy atom. Finally, MIRAS (Multiple Isomorphous Replacement with Anomalous Scattering) is used for multiple heavy atom derivatives collected at the fluorescence peak of the heavy atom. The formalisms for phase calculation are similar whether the signal is anomalous or isomorphous, as is the software that will allow you to exploit them.

11.3 Phases Combination

We have seen in these two paragraphs that it is possible to extract phase information from the positioning of a model or by using the signal from heavy atoms. But sometimes, after using one or the other, the uncertainty in the phases remains too large.

This can be observed by looking at the electron density map, which does not seem to be good enough to build, correct or complete a satisfactory atomic model.

In these cases, we can combine several sources of phase information, each of which is too weak on its own. These may be multiple sets of phases from heavy atoms with low phasing power, or from atomic models that are too partial or dissimilar to properly reflect the contents of the unit cell.

Without going into too much detail about the procedure used to combine these phases, it can be said that it consists of calculating the product of the individual phase distributions for each reflection, as defined in the next chapter. This procedure is used for any combination of phases, whether multiple isomorphic replacement (MIR) or the combination of molecular replacement phases with SAD, SIR or MIR phases.

Related videos

Experimental phasing: Isomorphous replacement (Phasage expérimental : le remplacement isomorphe)
https://youtu.be/bB7tYIf9WBM

Experimental phasing: Anomalous diffusion (Phasage expérimental : la diffusion anomale)
https://youtu.be/NdF70PTDAQU

Chapter 12

Phases Improvement and Model Building

The first phase estimates are often too inaccurate to be of use. A number of approaches can be used to improve them in order to obtain an interpretable electron density map.

12.1 Lack of Closure and Figure of Merit

When the structure is solved by molecular replacement, the first calculated phases come from the template model. This model can be more or less homologous or complete compared to the protein under study. Therefore, the error associated with the phases can be large, as can the bias introduced by the model (see figure 10.1). In the case of experimental phases, only the positions of the heavy atoms or anomalous scatterers are used to calculate the initial phases. The vector representation of the native data, the derived data and the structure factor calculated from the heavy atom position form a triangle (figure 11.2). However, the errors associated with the structure factor measurements or with the heavy atom positions and their occupancies are such that this vector triangle is not perfectly closed (figure 12.1). The errors associated with the phases can come from:

- the model of the heavy atom or of the anomalous scatterer (for each of them, we need to know the x, y, and z coordinates, the occupancy and the thermal motion factor, also called the temperature factor or the B factor);
- the amplitudes of the measured structure factors;
- the unit cell differences if the data come from several crystals (non-isomorphism).

In both cases, these errors lead to what is known as 'lack of closure'. For example, if we consider the three structure factors $F_P(hkl)$ of the native data, $F_{PH}(hkl)$ of the data containing a heavy atom, and $F_H(hkl)$ calculated from the coordinates of the heavy atoms, the triangle formed by the three vectors is not necessarily closed (figure 12.1). The value of ε joining the moduli of $F_{PH(\text{calc})}(hkl)$ (green) and $F_{PH(\text{obs})}(hkl)$ (blue) represents the lack of closure and provides an estimate of the error associated with the calculated phase.

DOI: 10.1051/978-2-7598-3501-0.c012

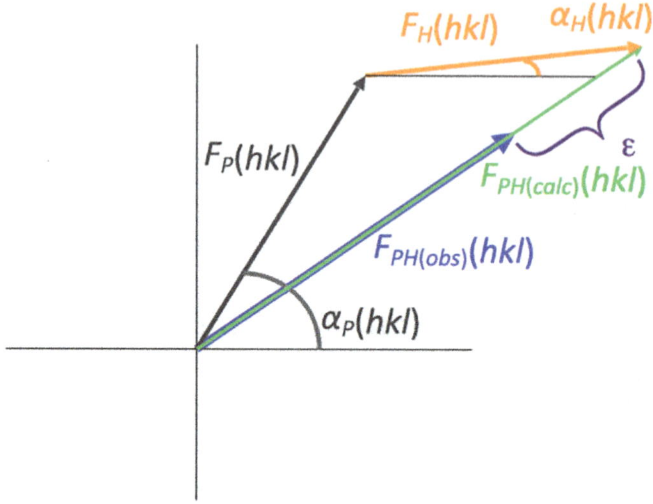

FIG. 12.1 – Vectorial representation of the native and derivative data and of the lack of closure ε.

We write:

$$\varepsilon = \left|F_{PH(\text{obs})}(hkl)\right| - \left|F_{PH(\text{calc})}(hkl)\right| \tag{12.1}$$

$\left|F_{PH(\text{obs})}(hkl)\right|$ is the modulus of the observed structure factor $F_{PH}(hkl)$;
$\left|F_{PH(\text{calc})}(hkl)\right|$ is the modulus of the calculated structure factor $F_{PH}(hkl)$.

The lack of closure can therefore be used to define a phase probability distribution, represented as the probability versus the possible value of the phase (figure 12.2). The distribution has a Gaussian shape and its centroid is called the best phase, while the highest value of the probability corresponds to the most probable phase. In the case of SIR phasing (chapter 11, figure 11.4, left), the phase ambiguity is reflected in the phase probability distribution by the presence of two maxima (figure 12.2, left). In the case of MIR phasing (chapter 11, figure 11.4, right), the most probable phase coincides with the best phase, and the ambiguity is resolved (figure 12.2, right).

The value of the best phase is the one used to calculate the electron density map. An average error $\Delta\alpha$ can be derived from the phase probability distribution which reflects the quality of the phases. The smaller $\Delta\alpha$, the better the phase estimate. The average value of $\Delta\alpha$ therefore provides a prior estimate of the quality of the electron density map that will be computed with these phases.

To simplify the interpretation, the value $m = \cos(\Delta\alpha)$ is used instead of $\Delta\alpha$. This is called the figure of merit.

When $m = 1$, $\Delta\alpha = 0$, there is no phase error.

When $m = 0$, $\Delta\alpha = 90°$, phases are equiprobable.

When $m = 0.5$, the phase error is 60°.

For example, a figure of merit of 0.23 is associated with a phase error of 76°, and a figure of merit of 0.96 is associated with a phase error of 16° (figure 12.2).

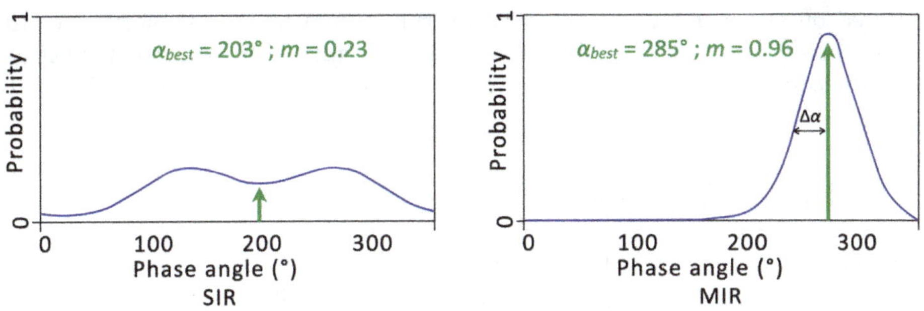

FIG. 12.2 – Representation of the phase probability distribution in the case of SIR (Single Isomorphous Replacement) on the left, or MIR (Multiple Isomorphous Replacement) on the right (adapted from 'The phase problem' (2003) G. Taylor. *Acta Cryst.* vol. D59, pp. 1881–1890). Note that the total area under each Gaussian must be identical for the two cases.

Regardless of the method used, the error associated with the first set of experimental phases is high. From the first value of the phases, it is possible to calculate a first electron density map. In order to minimise the weight of the errors in the calculation of this electron density map $\rho(xyz)$, we use the figure of merit m as a weighting coefficient for each amplitude. Thus, in the calculation of the electron density map, a reflection associated with a poor phase quality, *i.e.* with a figure of merit close to zero, has a small contribution. This first electron density map can be modified to improve the accuracy of the phases. The improvement in phase quality leads to an increase in the average figure of merit. We carry out successive iterations between the reciprocal space and the real space until we reach convergence, *i.e.* until the average figure of merit no longer changes significantly (figure 12.3).

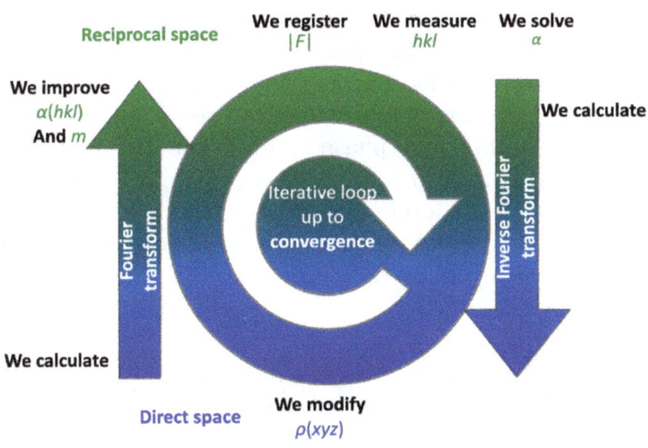

$$\rho(xyz) = 1/V \sum_h \sum_k \sum_l m \, |F(hkl)| \, e^{i\alpha(hkl)} \, e^{-2\pi i(hx+ky+lz)}$$

FIG. 12.3 – Overall iterative loop of phase improvement.

Based on successive iterations between the calculation of the electron density map in direct space and the improvement of the phases in reciprocal space, the process is continuous from the first electron density map to the evaluation of the final model.

12.2 Phases Improvement

Depending on the method used to solve the phases, electron density maps have different characteristics. The use of a model is part of the process of obtaining the phases in a molecular replacement approach. A model is therefore available from the first electron density map. The phase improvement consists of correcting the model. In the case of experimental phasing, the first step is to improve the electron density map in order to be able to interpret it, *i.e.* to construct the three-dimensional model of the electron density map.

12.2.1 Electron Density Modifications

There are many different statistical approaches to the process of phase improvement by electron density modification. They are based on the use of *a priori* knowledge of common crystal properties of biological macromolecules. To illustrate this step, we will use a representative modification, namely density flattening in the solvent regions.

A section of the electron density map shows dense, electron-rich regions with large intensity variations: these are the regions where the protein is located. Around these regions, the low-density regions correspond to the solvent channels. Since the solvent is disordered, the electron density in this region should be constant. The observed density peaks therefore reflect some of the errors inherent in phasing (figure 12.4, left). Solvent flattening involves first drawing a mask around the regions of dense and rapidly changing intensity and then assigning a constant value to $\rho(xyz)$ in the regions outside the mask, *i.e.*, the solvent regions (figure 12.4, centre). From this modified electron density map we can calculate a new set of structure factors and phases, which are the first improved phases associated with higher figures of merit. When combined with the experimental data, the new map, called the levelled map, shows a clear improvement, although there are still traces of density in the solvent regions due to residual errors associated with the phases (figure 12.4, right).

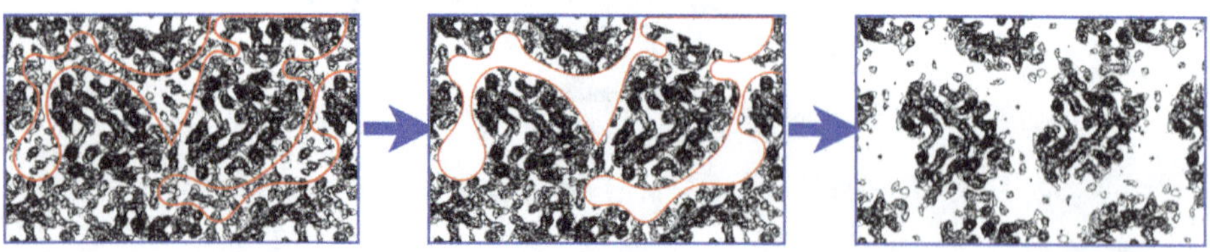

FIG. 12.4 – Phases improvement using solvent flattening.

As each structure factor contributes to the overall $\rho(xyz)$ map, and each $\rho(xyz)$ point contributes to the overall structure factors, we observe in particular that the quality of the electron density in the dense regions also improves. This improvement allows us to gradually build up the three-dimensional model of the molecule.

12.2.2 Iterative Loop of Phases Improvement

To facilitate the construction of the three-dimensional model, the aim is to obtain the best possible electron density map. An iterative approach is used (figure 12.5):

1/ The first step is to estimate the phase distribution and calculate the figure of merit m associated with each reflection.
2/ The first electron density map is calculated with the measured structure factors $F(hkl)$, the calculated phases and the figures of merit. The amplitudes $F(hkl)$ cannot be modified as they correspond to the experimental measurement.
3/ The electron density map is modified to refine the phase values.
4/ From a modified $\rho(xyz)$ map, we obtain a new set of phases calculated by inverse Fourier transform.
5/ The calculated data are combined with the experimental data to limit the bias introduced by the modification of the electron density map.

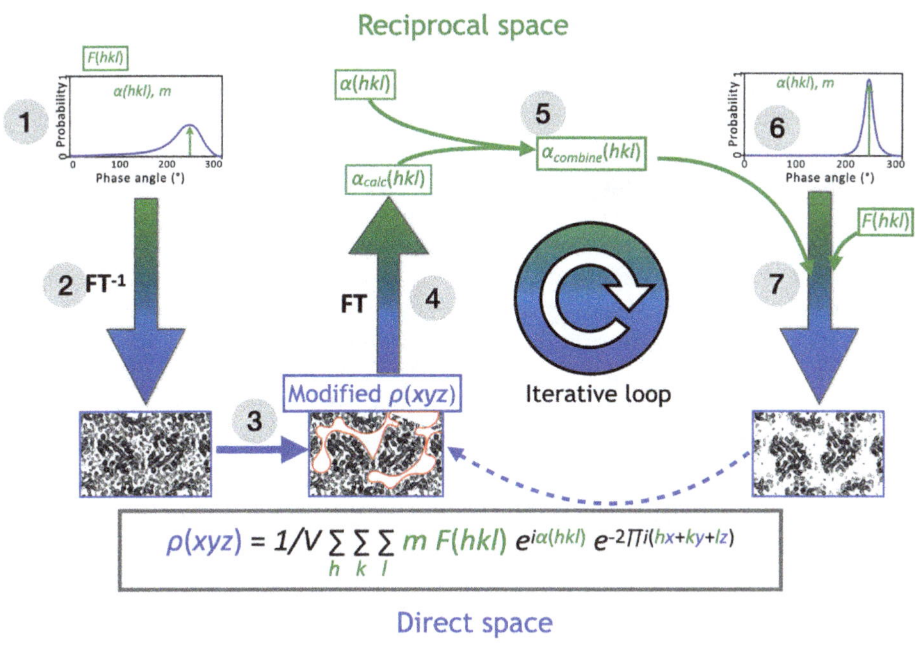

FIG. 12.5 – Steps of the iterative approach of phase improvement.

6/ If the modifications of the electron density map $\rho(xyz)$ are correct, the phases improve, the phase probability distribution becomes narrower and the figure of merit increases.
7/ The new electron density map computed with combined phases and new figures of merit is of better quality, allowing more accurate modifications.

Steps 4–7 are repeated iteratively until convergence is reached.

12.3 Molecular Model Building

Typical steps in molecular model building include:

- density skeletonisation;
- α-carbon chain construction;
- poly-alanine model construction;
- choice of chain direction;
- side-chain construction;
- more residues built?
    - o Yes -> new electron density map,
    - o No -> structure refinement.

For example, in the case of an electron density map, after phase improvement at a resolution of 2.5 Å, it is not possible to directly place atoms correctly (figure 12.6, A). It is necessary to proceed step by step:

- Skeletonisation consists of calculating the connectivity in the electron density map. The result is shown as a continuum line. The resulting skeleton is used to guide the subsequent steps (figure 12.6, B).
- In the example shown in figure 12.6, an alpha helix is clearly visible and, depending on the software, the corresponding carbon chain can be positioned automatically (figure 12.6, C). As the software does not allow us to choose between the two possible directions of the chain from the N-terminus to the C-terminus, we generally have two options at this stage.
- Some construction software will convert the carbon backbone directly into a poly-alanine polypeptide chain (figure 12.6, D). If the quality of the map allows, we can see that the correct orientation of the alpha helix from the N-terminus to the C-terminus is that which is compatible with the positioning of the side chains.

From a partial, *i.e.*, incomplete, model it is possible to calculate the structure factors F_{calc} and their phases. In the first stage of the construction process, the data calculated with the partial model are combined with the experimental data. Then, when the fraction of the constructed model is sufficient, only the phases calculated from the model are used.

The iterative loop described for phase refinement is similar to the construction phase. As the construction progresses, a transition can be made from construction to refinement.

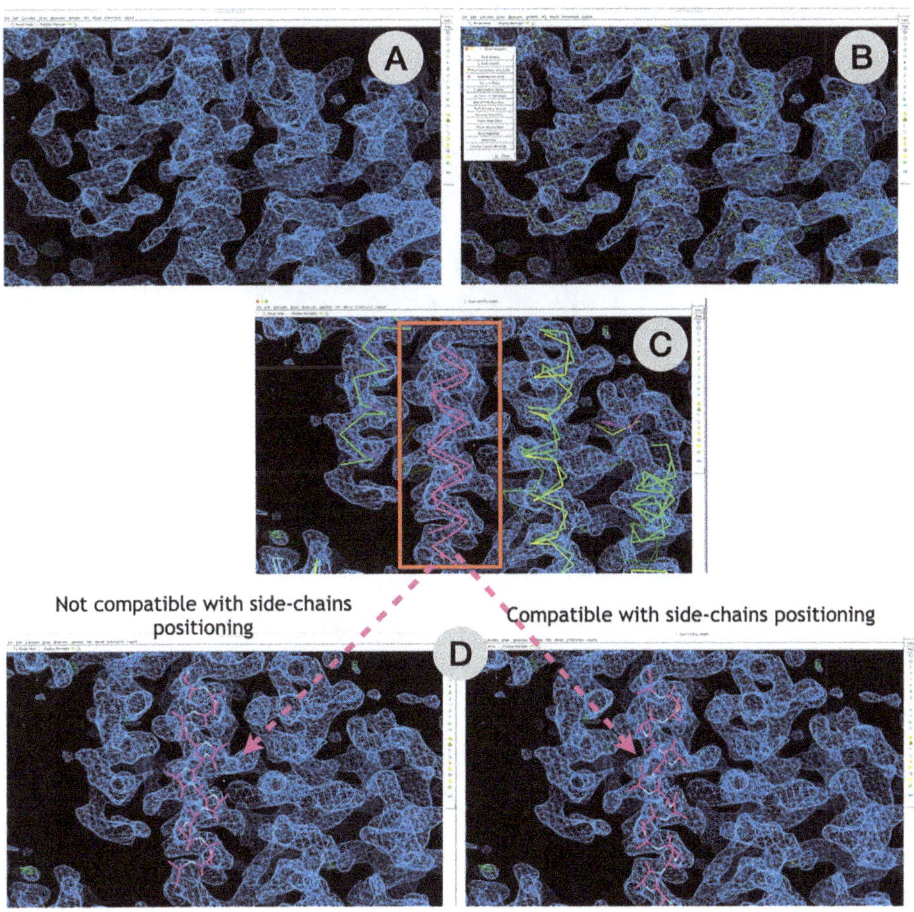

FIG. 12.6 – Example of the construction of a model region using the Coot software from CCP4 (https://www2.mrc-lmb.cam.ac.uk/personal/pemsley/coot/, and https://www.ccp4.ac.uk/), with the electron density alone in A, the skeletonisation in B, the construction of a-carbon chain and several α-helix and their possible orientation in C, the polypeptide chain of an α-helix in D in an orientation compatible (right) or not compatible (left) with side-chains positioning.

Related videos

Phases improvement and model building (Amélioration des phases et construction du modèle)
https://youtu.be/bB38X40dxTg

Chapter 13

Model Refinement and Evaluation

The aim of refining the model is to ensure that it best reflects the information contained in the diffraction data. It is an iterative process, alternating between so-called construction (or reconstruction) steps, which involve local corrections to the model in direct space, and so-called refinement steps, which aim at minimising the discrepancy between the calculated data and the experimental data in reciprocal space, using a set of geometric constraints applied to the molecular model.

13.1 Refinement

In reciprocal space, refinement consists of minimising the difference between the observed data, *i.e.* $F(hkl)_{\text{obs}}$ (or F_{obs} for short), and the data calculated from the three-dimensional model, *i.e.* $F(hkl)_{\text{calc}}$ (or F_{calc}). For the model derived from the interpretation of the electron density map, the aim is to reflect the molecular crystal structure as accurately as possible. The main parameters optimised during the refinement process are, for each atom, the x, y, z position coordinates and the B-factor, also known as the temperature factor, which reflects the thermal motion of the atoms around their position. Solvent molecules or ions bound to the macromolecule are not necessarily present in all cells. Therefore, we can also optimise their occupancy ratio, which can be less than 1. Finally, since the refinement is based on the difference between F_{obs} and F_{calc}, the scaling factor between the two must also be optimised.

During the refinement process, we monitor the improvement of the model through a reliability factor, the *R-factor*, which measures the average difference between F_{calc} and F_{obs}:

$$R = \frac{\sum_{hkl} \left| |F_{\text{obs}}| - |F_{\text{calc}}| \right|}{\sum_{hkl} |F_{\text{obs}}|} \tag{13.1}$$

The *R-factor* is comprised between 0 and 1. During the refinement process the model gradually improves, the gap between F_{obs} and F_{calc} decreases and the

DOI: 10.1051/978-2-7598-3501-0.c013

R-factor also decreases. The lower it is, the better the agreement between the model and the data. At the end of the refinement, it is generally between 0.15 and 0.25, depending on the quality and resolution of the data. However, the algorithms used by the refinement software directly seek to minimise this *R-factor*, which can lead to a possible bias in its use. In fact, the *R-factor* may decrease without any real improvement in the quality of the model. To avoid this bias problem, an *R-factor* is calculated using a small fraction of the reflections, between 3% and 10%, randomly selected at the beginning of the refinement process and systematically excluded from all refinement processes. These reflections are not 'seen' by these processes, they are independent of the refinement process. This *R-factor*, called *R-free*, is calculated in the same way as the *R-factor* and is used as a cross-validation procedure. In fact, refinement and reconstruction can be controlled by comparing R and *R-free*. At the end of the refinement, the *R-free* is generally between 0.20 and 0.30 and the difference between the *R-free* and the R is around 0.05 and should not exceed 0.1. Again, the resolution and quality of the experimental data will influence the *R-free* value. In general, the higher the resolution, the lower the R and *R-free* values, which reflects the better fit of the model.

Since the x, y, and z position of each atom changes during refinement, a second essential aspect is the geometry of the molecular model. A number of three-dimensional structures solved at very high resolution have made it possible to establish a standard geometry of biological macromolecules, described in libraries. To facilitate the use of these libraries, the geometric elements of macromolecules are described in terms of molecular blocks (see the box in the introduction on **The different types of macromolecules**). These blocks are amino acids in the case of proteins (see box in chapter 2 on **The 20 natural amino acids**), nucleotide bases in the case of nucleic acids or cofactors, and ligands. They are linked together to form, for example, polypeptide chains or nucleotide chains. The libraries also contain a detailed description of the geometric parameters of the possible bonds between these blocks. These are peptide bonds and disulphide bridges in the case of proteins, phosphodiester bonds in the case of nucleic acids, and so on. As the model evolves and the positions of the atoms change during refinement, we track the deviation between the geometry of the model and the geometric standards for bond lengths, angles, dihedral angles or improper angles (figure 13.1). The dihedral angle can be defined as the angle between two consecutive planes $\widehat{123}$ and $\widehat{234}$ (figure 13.1) and can be described as the 1-2-3-4 dihedral angle. Along the peptide chain, each amino acid participates in three successive bonds that are associated with as many dihedral angles, the angles ω, ϕ, ψ. Angle ω is the dihedral angle Cα-CO-NH-Cα, it corresponds to the peptide plane and its value varies very little around 180°. The angle ϕ is the dihedral angle CO-NH-Cα-CO and the angle ψ is the dihedral angle NH-Cα-CO-NH (figure 13.1). Their values are characteristic of the secondary structures of the proteins and we will see below that they are used to validate the model.

RMS deviations	
Bond length (Å)	0,010
Bond angles (°)	1,700
Dihedral angles (°)	24,70
Improper angles (°)	1,5

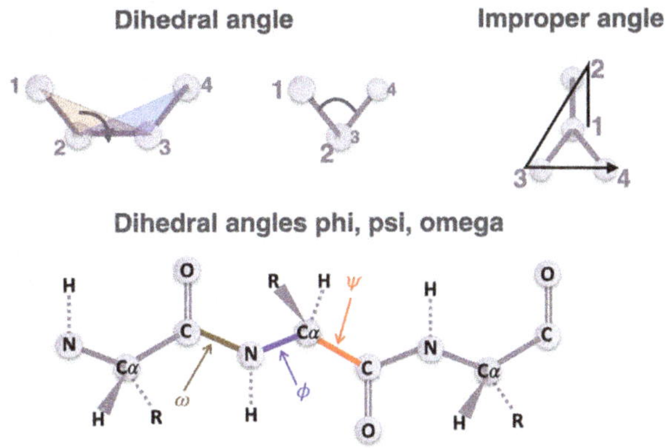

FIG. 13.1 – Standard deviations of geometric parameters from the ideal (left). Schematic representation of the different angles used to describe the geometry of a protein chain (right).

13.1.1 Iterative Process of Refinement

There are many refinement methods currently available, including classical least squares, molecular dynamics and maximum likelihood algorithms. We will not go into detail here. Typically, during a refinement run, the software is provided with:

- A coordinates file with x, y, z positions, temperature factors B and occupancy for each atom of the model. The $F_{\text{calc}}(hkl)$ are calculated using these parameters.
- The experimental data, the structure factors $F_{\text{obs}}(hkl)$.

After scaling of $F_{\text{calc}}(hkl)$ and $F_{\text{obs}}(hkl)$, the software modifies the molecular model both to minimise the deviation between $F_{\text{calc}}(hkl)$ and $F_{\text{obs}}(hkl)$ and to improve or constrain its geometric parameters. Thus during a refinement cycle, there is a constant back and forth between the reciprocal space and the direct space. At the end of a refinement cycle, we obtain:

- A new coordinates file with refined x, y, z positions, B factors and occupancies,
- A file containing $F_{\text{calc}}(hkl)$, α phases, and weighting factors as a figure of merit for each hkl index, in addition to the experimental data $F_{\text{obs}}(hkl)$.

Using these data, we calculate electron density maps and observe the refined model. This is where manual refinement, also known as reconstruction, begins. The model is examined in the electron density maps, corrected and completed manually. If the manual modifications improve the model, then the phases and the electron density map of the next cycle will also improve. Conversely, an error in the manual modifications leads to a degradation of the phases and the electron density map of the next cycle. The evolution of the *R-free* also controls this process. When sufficient modifications have been made, a new refinement cycle can be run (figure 13.2).

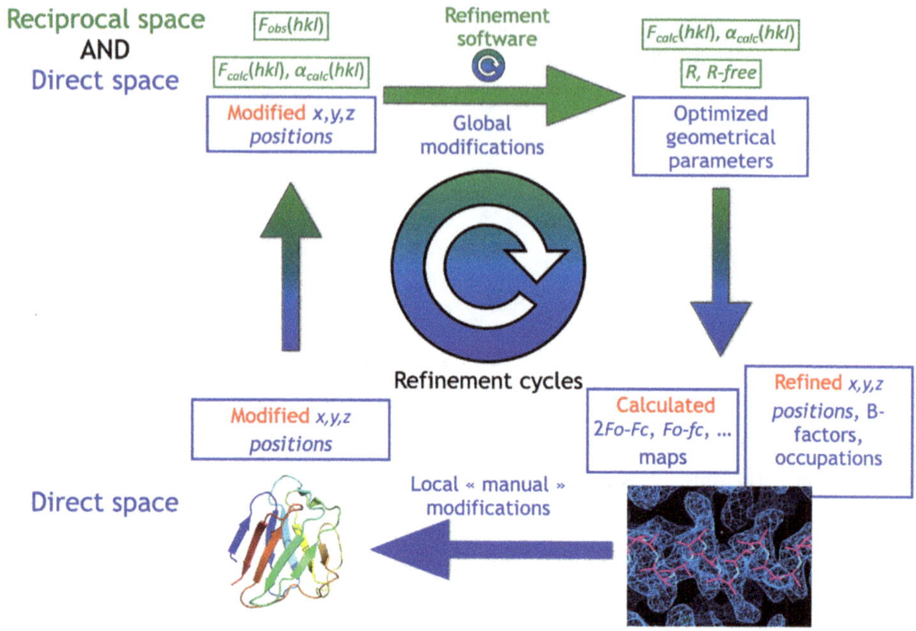

FIG. 13.2 – Refinement cycles.

13.1.2 The 2Fo-Fc and Fo-Fc Maps

At each model refinement cycle, new electron density maps are calculated. We have both the measured F_{obs} amplitudes and the F_{calc} amplitudes and α_{calc} phases calculated from the model (figure 13.2). The phase has a dominant weight in the Fourier series (see figures 8.6 and 10.1), so a map calculated using only F_{obs} and α_{calc} will be almost identical to the introduced model. It is biased by the model. To avoid this, in practice, Fourier difference maps are of the following type:

$$a \times m \times F_{\text{obs}} - b \times D \times F_{\text{calc}} \tag{13.2}$$

where a and b are integers,
m is the figure of merit,
D is a second weighting coefficient that takes into account the incompleteness of the model,
m and D are specific to each structure F_{calc}.

The most commonly used values for a and b are (2,1) for maps called 2*Fo-Fc* and (1,1) for maps called *Fo-Fc*. The 2*Fo-Fc* map contains an electron density at each point x, y, z which must correspond to that of the crystallised molecule (in blue in figure 13.3). It allows the electron density of the molecule to be observed and the model to be corrected. The *Fo-Fc* map contains peaks which are positive when parts of the molecular model are missing (green in figure 13.3), or negative when they correspond to over-modelled parts, *i.e.* a residue or an atom has been positioned where it should not be.

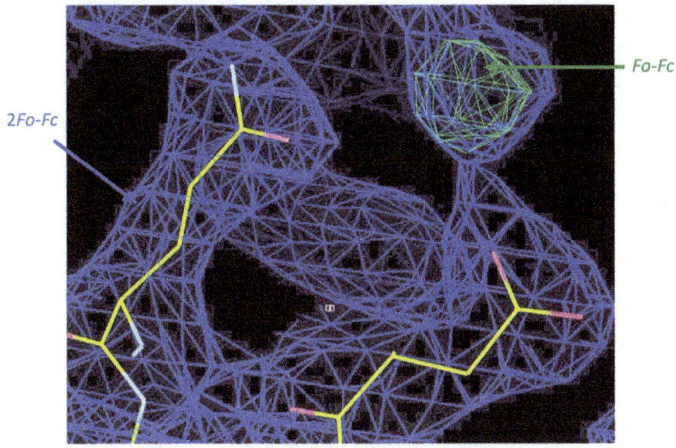

FIG. 13.3 – Electron density maps 2*Fo*-*Fc* (blue), *Fo*-*Fc* positive (green).

13.1.3 Refinement Limits

Regardless of the refinement method used, for a reliable result, the number of parameters that can be refined must not exceed the number of data available. This number of data, *i.e.* the number of F_{obs}, depends on:

- the data completeness;
- the maximum resolution of the data;
- the solvent content, which directly affects the amount of data for a given resolution.

Therefore, at low resolution, the number of experimental data is small and it is necessary to introduce constraints to limit the number of parameters to be refined and to keep the number of measured data higher than the number of parameters to be refined. On the contrary, at very high resolution we can relax the geometric constraints and refine more parameters (figure 13.4). Although this is not a strict procedure, depending on the maximum resolution of the experimental data, the refinement allows to refine:

- Above 3.8 Å: the x, y, z coordinates of residues or domains by rigid blocks, the geometry is constrained by domain.
- Between 3.8 and 3.2 Å resolution: individual x, y, z atomic coordinates, and groups of temperature factors, generally grouped by globular domains. The weight of geometric constraints is high, which means that there is little variation around the expected values.
- Between 3.2 and 2.7 Å resolution: individual x, y, z, individual temperature factors, Translation-Libration-Screw-Rotation factors (called TLS), and few stable solvent molecules.
- Between 2.7 and 2.0 Å resolution: individual x, y, z, individual temperature factors, Translation-Libration-Screw-Rotation factors (called TLS), as many solvent molecules as the number of residues in the molecule.

- Below 2.0 Å: individual x, y, z, several alternate conformations, anisotropic individual temperature factors, TLS, a number of solvent molecules equivalent to 1.5 times the number of residues in the molecule.
- Below 1.2 Å: we start to see the hydrogens of the molecule.

Resolution	Refinement
> 3.8 Å	Rigid blocks coordinates
3.2 - 3.8 Å	Atomic coordinates *x, y, z* Temperature factors per domain
2.7 - 3.2 Å	Atomic coordinates *x, y, z* Individual temperature factors *B* Translation-Libration-Screw-Rotation (TLS) Some stable solvent molecules
2.0 - 2.7 Å	*x, y, z* *B* individuals TLS Solvent molecules
≤ 2.0 Å	*x, y, z* + alternate conformations Anisotrope individual *B* TLS Solvent molecules
≤ 1.2 Å	We start adding hydrogens

FIG. 13.4 – Summary of parameters that can be refined depending on the resolution, *i.e.* depending on the amount of experimental data.

Let's take the helix from the previous chapter (figure 12.6) and look at it in electron density maps at different resolutions to illustrate the problem of data number and resolution (figure 13.5):

- At 5 Å resolution, we can just position the α-helix per rigid block.
- At 4 Å resolution, the main chain of the helix is placed.
- At 3.5 Å resolution, we properly orientate the α-helix as well as a few side chains.
- At 3 Å resolution, the side chains are placed without ambiguity.
- At 2.5 Å resolution, all the side-chains are positioned.

Depending on the biological information required, a structure with a resolution of 5 Å is still interesting, for example, to determine the general shape of a protein or the global interaction interface between two macromolecules. A resolution of 2.5 Å or more is generally the optimal resolution in biology, for example, to describe interaction networks.

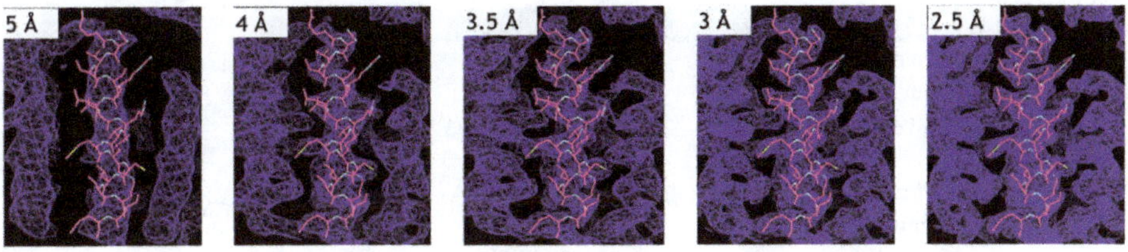

FIG. 13.5 – Refinement level according to data resolution.

13.2 Three-Dimensional Structure Validation

When the *R-factor* is between 0.15 and 0.25, the *R-free* between 0.20 and 0.35, and they no longer evolve, and the *Fo-Fc* map no longer shows significant peaks of positive density, then the convergence of the refinement has been reached and it is time to evaluate the structure and complete the statistical table that will accompany the structure and the scientific publication. This table compiles a set of crystallographic and geometric parameters that evaluate the agreement between the atomic model and the experimental data from which it was derived, as well as the quality of the atomic model in comparison with the standards established by the scientific community. It should be noted that for some of these criteria a 'tolerance' is allowed to take into account the maximum resolution of the crystallographic data from which the model was constructed. The first part of the table gives information on the experimental data and their acquisition conditions: the wavelength at which they were recorded, the space group, the cell parameters, the diffraction limits, the agreement factor of the diffraction data, the number of reflections, the signal-to-noise ratio, the completeness of the data and their redundancy. For each parameter, the value in brackets refers to the same parameter for the highest resolution range. The second part of the table shows the statistics related to the quality of the model. In the case of molecular replacement, it is generally the values associated with the refinement of the model that validate the molecular replacement solution. These include resolution, *R-factor*, *R-free*, number of atoms included in the model, number of water molecules, average figure of merit, geometric deviations from reference values the code assigned by the PDB when the model was deposited and the structure factors that must be obtained before publication (figure 13.6).

In the case of experimental phasing, *e.g.* by the MAD method with data from different crystals and different space groups, the statistics table contains more information (figure 13.7).

Independently of the experimental data, the geometric parameters of the model make it possible to assess its quality. Among the criteria that are frequently used for proteins, the Ramachandran diagram, developed in 1963 by G.N. Ramachandran, makes it possible to visualise the values of the dihedral angles ϕ and ψ (figure 13.1), and to check that they are located in areas compatible with the secondary structures of proteins (figure 13.8).

Data collection		
Wavelength (Å)	0.98	
Space-group	*P2₁*	
Unit cell : a, b, c (Å) α, β, γ (°)	53.8, 104.8, 85.3, 90, 94.4, 90	
Diffraction limits (last shell)	2.44 Å (2.63 Å – 2.44 Å)	Data
R-meas	0.068 (0.664)	
CC(1/2)	99.9 (54.8)	
Number of unique reflections	33979 (5718)	
I/σ	14.29 (2.26)	
Completeness	0.965 (0.794)	
Redundancy	7.8 (7.5)	
Refinement		
Resolution	2.44 Å	
R-factor	0.193	
R-free	0.237	
Number of protein atoms	6938	
Number of water molecules	94	Model
Figure of merit	0.7813	
RMSD bond lengths (Å)	0.004	
RMSD bond angles (°)	0.976	
PDB entry	4RQI	

FIG. 13.6 – Example of a table of statistics in the case of a structure solved by molecular replacement.

Data collection	MAD		Native	
	Remote	Inflection	Cubic	Monoclinic
Wavelength (Å)	0.9762	1.0087	0.98	0.98
Space-group	I432			$P2_1$
Unit cell : a, b, c (Å) α, β, γ (°)	a=b=c=187.73 Å α =β=γ=90°			a=106.9 b=136.8 c=140.5 α=γ=90° β=97.3°
Diffraction limits (last shell)	99-3.5 (3.66-3.5)	99-3.55 (3.68-3.55)	50-3.15 (3.22-3.15)	40-2.7 (2.76-2.7)
R-meas	0.097 (0.858)	0.093 (0.801)	0.071 (0.943)	0.09 (0.975)
CC(1/2)	99.9 (61.2)	99.8 (57.3)	99.9 (56.6)	99.9 (73.1)
Number of unique	8068	7136	7994	110121
I/σ	32.2 (1.2)	34.8 (1.3)	11 (1.09)	12.4 (1.6)
Completeness	100 %	100 %	98 %	100 %
Redundancy	>13	>13	3	3
Phasing				
Average figure of merit	0.3724			
Refinement				
R-factor				0.245
R-free				0.315
Total number of atoms				18548
Number of :	Protein atoms			15976
	Carbohydrate atoms			1718
	Sulfate ions atoms			105
	Water molecules			749
RMSD bond lengths (Å)				0.011
RMSD bond angles (°)				1.634
PDB entry				1YWH

FIG. 13.7 – Example of a statistics table in the case of a structure solved by the MAD method, with several different crystals.

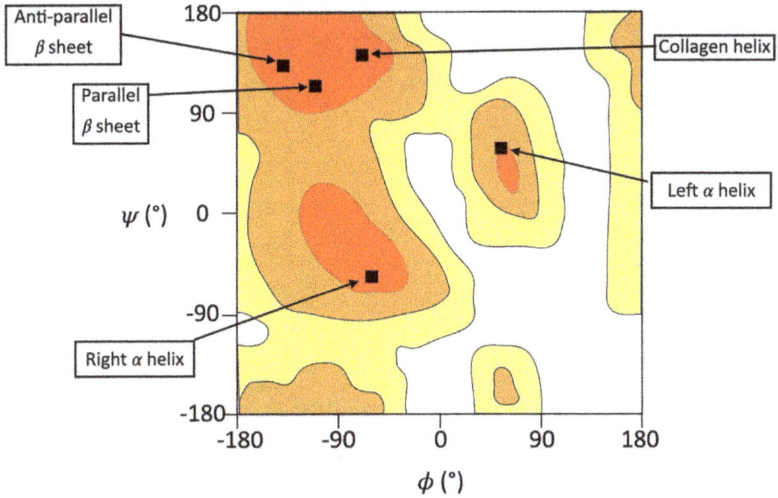

FIG. 13.8 – Ramachandran diagram with regions corresponding to the main secondary structures in the most favoured regions in red, additional permitted regions in orange, generously permitted regions in yellow, and forbidden regions in white.

Finally, the PDB file itself contains a lot of information, and for newer structures, a reflection file is associated with the PDB code in addition to the coordinate file. Thus, the coordinates in the PDB file allow the calculation of F_{calc} and phases, and the reflection file provides F_{obs}. It is therefore possible to calculate an electron density map to observe and evaluate the three-dimensional structure of interest.

Related videos

Model refinement and evaluation (Affinement, interprétation et évaluation du modèles)
https://youtu.be/BpSx_vsJnak

To Go Further

Some books:

- Biochemistry (ninth edition) Jeremy M. Berg, Lubert Stryer, John Tymoczko, Gregory Gatto. Edition MACMILLAN (2019). ISBN 10: 1319114652/ISBN 13: 9781319114657
- Lehninger Principles of Biochemistry: International Edition (7th edition). David L. Nelson, Michael M. Cox. Edition W.H. Freeman & Co Ltd (2017). ISBN-10: 9781319108243/ISBN-13: 978-1319108243
- The Biophysical Chemistry of Nucleic Acids & Proteins. Thomas E. Creighton. Edition Helvetian Press, 2010. ISBN 10: 0956478115/ISBN 13: 9780956478115
- Crystallography Made Crystal Clear: A Guide for Users of Macromolecular Models (Complementary Science) (English Edition). Gale Rhodes. Academic Press; Edition: 3rd edition (16 février 2006). Collection: Complementary Science. ISBN-10: 0125870736/ISBN-13: 978-0125870733
- Biomolecular Crystallography: Principles, Practice, and Application to Structural Biology. Bernhard Rupp. Edition Garland Science (2009). ISBN-10: 0815340818/ ISBN-13: 978-0815340812
- International Tables for crystallography, ISBN: 978-0-470-68575-4 doi: https://doi.org/10.1107/97809553602060000001, volumes A to H, https://it.iucr.org/
- Principles of proteins X-ray crystallography, Jan Drenth, ISBN-13: 978-0-387-33334-2
- Fundamentals of crystallography (3rd edn), Giacovazzo, online ISBN: 9780191775000

Some websites:

- CCP4: https://www.ccp4.ac.uk/
- Protein Data Bank: https://www.rcsb.org/
- European web of the Protein Data Bank: https://www.ebi.ac.uk/pdbe/

- Absorption edge for elements of the periodic table: http://skuld.bmsc.washington.edu/scatter/AS_periodic.html

Some of the founding papers:

- The detection of sub-units within the crystallographic asymmetric unit (1962) M.G. Rossmann and D.M. Blow. *Acta Cryst.* vol. 15, pp. 24–31. DOI: https://doi.org/10.1107/S0365110X62000067
- Determination of macromolecular structures from anomalous diffraction of synchrotron radiation (1991) W.A. Hendrickson. *Science* vol. 254, pp. 51–58. DOI: https://doi.org/10.1126/science.1925561
- Sparse matrix sampling: a screening method for crystallization of proteins (1991) J. Jancarik and S.-H. Kim. *J. Appl. Cryst.* vol. 24, pp. 409–411. DOI: https://doi.org/10.1107/S0021889891004430
- Phase problem in X-ray crystallography, and its solution (2001) K. Cowtan. *Encyclopedia of life science*, Macmillan Publishers Ltd, Nature Publishing Group. https://people.bu.edu/mfk/restricted566/phaseproblem.pdf
- The phase problem (2003) G. Taylor. *Acta Cryst.* vol. D59, pp. 1881–1890. DOI: https://doi.org/10.1107/S0907444903017815
- An introduction to molecular replacement (2008) P. Evans and A. McCoy. *Acta Cryst.* vol. D64, 1–10.

www.ingramcontent.com/pod-product-compliance
Lightning Source LLC
Chambersburg PA
CBHW071345230926
51846CB00002B/454

* 9 7 8 2 7 5 9 8 3 5 0 1 0 *